A. Desmond O'Rourke, PhD
Editor

Understanding the Japanese Food and Agrimarket: A Multifaceted Opportunity

Pre-publication
REVIEWS,
COMMENTARIES,
EVALUATIONS . . .

"The book is a major contribution in that it brings together an enormous amount of important information in a logical fashion on marketing agricultural commodities and products in Japan from a U.S. perspective. THE BOOK GOES A LONG WAY IN CUTTING THROUGH THE MYSTERIES OF MARKETING IN A VERY COMPLEX JAPANESE SOCIETY. Primarily because of my research interests, the most informative sections of the book for me were those dealing with Japanese decision-making, the food distribution system, and the demand for fruits and vegetables. Moreover, I will draw heavily from the book in teaching agribusiness and marketing."

James E. Epperson, PhD
Professor of Agriculture
and Applied Economics,
The University of Georgia

Understanding the Japanese Food and Agrimarket

A Multifaceted Opportunity

FPP Agricultural Commodity Economics,
Distribution, & Marketing
A. Desmond O'Rourke, PhD
Senior Editor

New, Recent, and Forthcoming Titles:

Marketing Livestock and Meat by William Lesser

Understanding the Japanese Food and Agrimarket: A Multifaceted Opportunity edited by A. Desmond O'Rourke

The World Apple Market by A. Desmond O'Rourke

Understanding the Japanese Food and Agrimarket

A Multifaceted Opportunity

A. Desmond O'Rourke, PhD
Editor

CRC Press
Taylor & Francis Group
Boca Raton London New York

CRC Press is an imprint of the
Taylor & Francis Group, an **informa** business

CRC Press
Taylor & Francis Group
6000 Broken Sound Parkway NW, Suite 300
Boca Raton, FL 33487-2742

First issued in paperback 2019

© 1994 by Taylor & Francis Group, LLC
CRC Press is an imprint of Taylor & Francis Group, an Informa business

No claim to original U.S. Government works

ISBN-13: 978-0-367-40210-5

Library of Congress Cataloging-in-Publication Data

Understanding the Japanese food and agrimarket : a multifaceted opportunity / A. Desmond O'Rourke, editor.
    p.   cm.
    Papers from a conference held to note the 5th anniversary of the IMPACT Center of Washington State University.
    Includes bibliographical references and index.

    1. Food industry and trade–Japan–Congresses. 2. Produce trade–Japan–Congresses. 3. United States–Commerce–Japan–Congresses. 4. Japan–Commerce–United States–Congresses. I. O'Rourke, A. Desmond (Andrew Desmond), 1938– . II. Washington State University. IMPACT Center.
HD9017.J42U516   1993
381'.45664'00952–dc20   92-20076
                                                    CIP

Visit the Taylor & Francis Web site at
http://www.taylorandfrancis.com

and the CRC Press Web site at
http://www.crcpress.com

CONTENTS

ABOUT THE EDITOR

Andrew Desmond O'Rourke, PhD, is Professor of Agricultural Economics, and Director of the International Marketing Program for Agricultural Commodities and Trade at Washington State University in Pullman, Washington. Dr. O'Rourke has been involved in research, teaching, and consulting in the domestic and international aspects of marketing since 1960 in both the public and private sectors. He is the author of two textbooks, a contributor to a number of other books, and the author or co-author of over 150 journal articles, research bulletins, and popular articles on marketing topics. His areas of interest have included marketing of North American fruits, beef, dairy products, seafood, vegetables, and grains in all five continents. Since 1985, Dr. O'Rourke has been Director of the IMPACT Center, an integral part of the College of Agriculture and Home Economics at Washington State University. The IMPACT Center is a multidisciplinary program which applies science and technology to the solution of international agricultural marketing problems.

ABOUT THE CONTRIBUTORS

Suzanne Callender, MA, was a graduate research assistant in agricultural economics at Washington State University, working on a multidisciplinary study of the market for juice, wine, and grapes in Japan.

Annabel Kirschner Cook, PhD, is an Extension sociologist in the Department of Rural Sociology at Washington State University. She has studied the impact of demographic and social changes on food demand in many Asian countries.

Bill B. Dean, PhD, is an IMPACT (International Marketing Program for Agricultural Commodities and Trade) Center Extension specialist and professor in the Department of Horticulture and Landscape Architecture at Washington State University. He has extensive experience in the potato and vegetable industries and has worked on improving the harvesting, handling, and shipping of fresh vegetables to Japan.

Raymond J. Folwell, PhD, is a professor of agricultural economics at Washington State University. He has conducted numerous studies of grape and wine markets for over two decades.

Joshua A. Greenberg, PhD, was a graduate research assistant in agricultural economics at Washington State University. He is currently an assistant professor of Natural Resource Economics in the School of Land and Resource Management at the University of Alaska at Fairbanks.

Manfred Heim, MA, a research associate in agricultural economics at the IMPACT Center at Washington State University, has conducted a number of studies on the demand for various fruits and vegetables in Japan.

Eric Hurlburt, MSc, is a program manager with the Market Development Division of the Washington State Department of Agriculture. He is responsible for agricultural trade issues arising in Japan, Canada, and Europe and for seafoods and specialty crops.

Kristen A. Johnson, PhD, is an assistant professor of animal science at Washington State University, specializing in beef cattle nutrition.

Raymond A. Jussaume, Jr., PhD, is an assistant professor in the Department of Rural Sociology at Washington State University. He has conducted extensive research in Japan.

John Konovsky, MSc, was a graduate research assistant in agronomy at Washington State University and was involved in the transfer of edamame technology from Japanese to Pacific Northwest growers. He is currently employed in the private sector.

Thomas A. Lumpkin, PhD, is an associate professor of agronomy at Washington State University, specializing in alternative crops. A regular visitor to Japan and China, he has been a leader in adapting native Asian foods for production and processing in the United States.

Thomas M. Maloney, MEng, is director of the Wood Materials and Engineering Laboratory at Washington State University. He has pioneered a number of innovative programs to enhance international trade in wood and wood products.

Scott C. Matulich, PhD, is a professor of agricultural economics at Washington State University. He has conducted numerous studies of the Alaskan fisheries and of the role of Japanese investors, fishermen, and buyers.

Dean McClary, MSc, is a graduate research assistant in agronomy at Washington State University. He has been heavily involved in the transfer of alternative crop technology from Japan to the Pacific Northwest, primarily for Azuki beans.

Ron C. Mittelhammer, PhD, is a professor of agricultural economics at Washington State University, specializing in econometric analysis.

Danna Moore, PhD, was a research associate in agricultural economics on special assignment with the IMPACT Center at Washington State University. She has completed a major survey of the potential for Hard White Wheat in major Asian markets, including Japan.

A. Desmond O'Rourke, PhD, is a professor of agricultural economics and director of the IMPACT Center at Washington State University. The IMPACT program specializes in bringing science and technology to bear on the solution of international marketing problems.

Roy F. Pellerin, MSc, is a professor of civil and environmental engineering and leader of the nondestructive testing program in the Wood Materials and Engineering Laboratory at Washington State University.

Yeshajahu Pomeranz, PhD, is leader of the IMPACT Center program in cereal chemistry and a professor of food science and human nutrition at Washington State University. He is recognized as an international authority in the field of cereal chemistry.

Dorothy Z. Price, PhD, is a professor in the Department of Child, Consumer, and Family Studies at Washington State University. She has conducted numerous comparative studies of the processes of decision making in the United States and in other countries.

Jerry J. Reeves, PhD, is a professor of animal science at Washington State University and co-leader of the project to establish the Japanese breed of cattle in the Pacific Northwest.

R. Thomas Schotzko, PhD, is an Extension specialist and professor of agricultural economics at Washington State University. He has focused on problems in handling and marketing fresh fruits and vegetables in the United States and abroad.

Raymond W. Wright, Jr., PhD, is a professor of animal science at Washington State University, noted for his work on animal fertility. He and his colleagues have led the effort to establish the Japanese Wagyu breed of cattle in the Pacific Northwest.

David Youmans, PhD, is an IMPACT Extension trade specialist at Washington State University, with primary responsibility for grains and livestock. He has lived and worked in many different countries. His major educational focus has been in helping Extension agents and their clients to understand the changing needs of key markets such as Japan.

Chapter 1

Introduction

A. Desmond O'Rourke

Despite the importance of the Japanese market to U.S. agricultural exporters, systematic monitoring and assessment of that market has been limited. Much of what we think we know about the Japanese market is based on stereotypes developed from the casual observations of short-term visitors. These generalities are of little value to the marketer of a specific agricultural product. They also ignore the dynamic ways in which world and Japanese society are changing.

The IMPACT (International Marketing Program for Agricultural Commodities and Trade) Center at Washington State University was set up in 1985 to provide Pacific Northwest agriculture with systematic and continuing information on export markets. From the beginning, our primary focus has been on the Japanese market. Many disciplines have been tapped to help us better understand that market, including agronomy, agricultural economics, business, engineering, foreign languages, horticulture, animal science, and food science.

On our fifth anniversary, in 1990, we felt it appropriate to organize a conference to take stock of what scientists in these different disciplines had taught us about the Japanese market for agricultural and forest products. We also hoped that this stocktaking would help us in planning more effectively for future studies of the Japanese market.

The first session of the conference focused on understanding Japanese consumers and the food distribution system. Consumer economist Dr. Dorothy Z. Price stressed the characteristic differences in both business and consumer decision making between the

United States and Japan. The Japanese have a much greater tendency to make decisions by consensus. Parties affected by a decision tend to be included in discussions that formulate the decision. In doing business with Japanese companies, the American exporter must be patient.

Rural sociologist Dr. Annabel Kirschner Cook pointed out that while the proportion of women working in Japan has risen steadily, Japanese women tend to drop out of the labor force while their children are being raised. Caring for children, daily shopping, and a preference for fresh product purchases are interrelated phenomena. However, Japan's fertility level (1.4 children per adult female) is now well below replacement levels. The U.S. fertility level is 1.9, and the U.S. permits immigration. So, Japan faces some wrenching demographic trends as the proportion of its population over the age of 45 rises. Japanese food consumption patterns are reflecting, and will reflect, the growth in influence of this more traditional segment of the population.

Another rural sociologist, Dr. Raymond A. Jussaume, Jr., stressed that Japan has not one but many distinct distribution systems. Even in beef, there are separate systems for Black Wagyu, for Red Wagyu, and for other beef cattle. In the fruit juice market, some products are handled by beer companies and some by beverage makers as an extension of their main product lines. Many retailers carry only one company's products, so site control is very important. Price tends to be fixed, so competition is based on branding and advertising. The gift market is important to many distribution channels, and direct distribution of products is growing as an alternative channel.

Eric Hurlburt, of the Washington State Department of Agriculture, described how state government works to stimulate trade in agricultural products. Japan is particularly important to the state of Washington as a market for 43 percent of its agricultural exports and 70 percent of its seafood sales. The Washington State Department of Agriculture supports a full-time representative in Tokyo as well as a number of trade specialists stateside. This setup acts as a funnel for information between Washington companies seeking business in Japan and Japanese companies seeking either product supplies or investment opportunities in the state of Washington.

State governments, in conjunction with the federal government, can also be critical in negotiating with Japanese bureaucracies on access for U.S. products. The Tokyo office can provide on-the-spot market analysis. Hurlburt reemphasized the need for patience and persistence in penetrating the Japanese market, the importance of coordinating U.S. efforts in every phase of marketing, and the need for long-term commitments by state and federal governments if they wish their efforts to be taken seriously in Japan.

A second session focused on understanding the marketing system for major agricultural commodities. Dr. David Youmans described the efforts of Cooperative Extension to prepare the Washington beef industry for liberalization of the Japanese beef market. An intensive market study in Japan was used to develop a training publication, video, and numerous presentations. Documentable improvements in production, feeding, packing, slaughter, etc., have been made as a result. The old way of doing beef business in Japan has been changed irreversibly, and more surprises are expected in the wake of full liberalization in April 1991.

Dr. Bill B. Dean contrasted the importance of vegetables in the Japanese diet with their rather poor reputation among Americans. The Japanese produce most of their vegetable needs. The U.S. is seen as just one of many supplementary sources. Washington has been the most competitive in processed vegetables, but there is also good potential for cauliflower, lettuce, onions, pumpkins, spinach, and vegetable seed.

Economist Dr. R. Thomas Schotzko suggested that the Japanese apple market is not a single, homogeneous market. Many of its common practices differ from those customary in Washington. The Japanese are much more concerned about soluble solids. They do not wax their apples. Carton size is based on the number of 5-kilogram trays included. Grade standards are different, and most of their apples are round. The Washington apple industry must determine how much it will need to adapt its practices to market effectively in Japan.

The last talk of the session, by Suzanne Callender, focused on the Japanese wine market. Imported European wines have set the standard for taste, quality, and prestige in the Japanese market. U.S. wine, in total, is still struggling to maintain its niche. Washington

State has an opportunity to promote itself as a premium-quality wine region. However, because its firms are too small to mount individual campaigns, there is a need for a collaborative effort to market Washington as a premium wine region.

In a third session, the focus shifted to demand for various key commodities in Japan. Dr. Danna Moore described the changes that have taken place in the Japanese wheat market as demand for white bread has declined and demand for specialty breads, cakes and confectioneries, and Chinese noodles has grown. A proposed U.S. wheat variety, Hard White Wheat, has potential because of its end-use properties. However, government controls on the Japanese grain importing system must be removed before millers and bakers can buy without political constraints. This could happen as the result of future General Agreement on Tariffs and Trade (GATT) negotiations.

Dr. Yeshajahu Pomeranz pointed out that most of what the Japanese learned about milling and baking they learned from Americans. However, they have now made their own breakthroughs in science and application and are ahead of us in development of new equipment, in developments in shortening, and in their attitude toward wheat enhancers and improvers. In the future, they will be looking more and more at products that provide convenience, high nutrition, low calories, and unique fibers and that are also micro-waveable.

Agricultural economist Manfred Heim pointed out that the Japanese apple market may be a difficult one for Washington exporters for a number of reasons. Per capita consumption of all fresh fruits, including apples, is stagnant. The Japanese have switched heavily to Fuji, Mutsu, and Tsugaru and away from Starking Delicious and Golden Delicious varieties. However, Washington Red Delicious differ in appearance and quality from Japanese Delicious and may be used to tap a separate market niche. There may also be an opportunity for mid-season sales from controlled atmosphere storage. Like Dr. Schotzko, Heim stressed the need for the Washington apple industry to investigate the Japanese market more thoroughly in anticipation of being allowed to enter that market.

Dr. Scott C. Matulich highlighted the unusual situation developing in the Seattle-based Alaskan king crab industry. Many Japanese companies have a financial stake in U.S. firms. As early-season cash buyers, they provide much of the industry's cash flow. They buy raw frozen sections that are remanufactured into many high-value products. In contrast, the U.S. market is for cooked frozen sections sold as a main dish. As prices rise, U.S. demand may dry up. It is expected that 90 percent of Alaskan king crab will be shipped to Japan this year. Similar outcomes are possible in other natural resources where Japan is the major export market.

A final session looked at new products and product adaptations being made to satisfy Japanese market needs. Roy F. Pellerin, of the Wood Materials and Engineering Laboratory, reported on collaborative efforts with key Japanese scientists to establish a common basis for U.S. and Japanese lumber grades, standards, and building codes (all of which have often been nontariff barriers in the past). He and his colleagues have been assisting the Japanese in applying nondestructive testing methods to their post-war forest lumber. This, in turn, is giving the Japanese an appreciation of the quality characteristics of Washington wood products.

Dr. Thomas A. Lumpkin reported on two new products being developed for the Japanese market. Plantings of Red Azuki beans in Washington have gone from zero to 325 acres in three years. Growers in 1990 will gross about $1000 per acre, with yields of about one ton; commercial production began in 1991. A newer project on edamame (edible soybeans) also shows promise. Yields have been about two tons per acre. A number of commercial cooperators have been found. Quality and processing time from Washington appear very competitive. There also seems to be a large market for this product in the U.S. A number of other alternative crops–such as wasabi radish, sprouted garlic, scarlet beans, sweet corn, and fava beans–also show promise for the Japanese market.

Dr. Raymond W. Wright, Jr. described the program to build a Wagyu cattle industry from scratch in the Pacific Northwest. After scrambling to get breeding stock, semen, and embryos, there is now a small herd in Pullman, Washington. Calves will also be raised cooperatively with private producers, so extensive research can be conducted on evaluating carcasses, feeding trials, and progeny testing. The goal of the project is to tap directly into the lucrative Wagyu beef market in Japan. However, preliminary tests suggest that the Wagyu may have a healthier quality of fat, which will make it desirable even in the U.S. market.

The conference reconfirmed that there is no substitute for detailed, accurate, and up-to-date information on the product market in Japan in which one is interested. The exporting firm needs to know the specifics of consumer demand, pricing practices, market channels, competition, promotional opportunities, packaging requirements, etc., relating to its product. Scientists need to intensify their efforts to get that specific information and to make it available to exporters and potential exporters.

The Japanese market is not a single monolithic entity. It shows many differences between regions, age groups, and market channels. It is buffeted both by worldwide events such as changing energy availability and by internal changes such as its low birthrate and rapidly aging population.

Depending on the product, the Japanese market may be wide open, offer occasional or seasonal opportunities, or be virtually impenetrable. In any case, successful exporting to Japan requires persistence in gaining access, rigorous control of quality requested by buyers, and a long-term commitment to servicing the market. For those firms that can meet exacting standards, the Japanese market can be very rewarding.

Chapter 2

Changes in Japanese Food Consumption

Annabel Kirschner Cook

INTRODUCTION

With the rapid modernization of Japan, some writers contend that Japanese consumption patterns are not just changing but becoming "Westernized." It is true that Japanese society has undergone a number of changes that influence consumption patterns. In the last twenty years, real income in Japan has increased dramatically. Generally as incomes increase, families spend a smaller percentage of the overall household budget on food, because they can afford to buy more discretionary items (Engel's Law). Also, the kinds of food consumed change as diets shift from a heavy reliance on grains and starches to a greater consumption of proteins (Bennett's Law) (Poleman, 1981). In addition, Japan has urbanized rapidly, and population growth continues to be concentrated in the country's major urban areas. There are also increasing numbers of Japanese women in the labor force. In the U.S. these trends have been associated with the greater use of convenience and prepared foods, and with increased eating out.

However, we should be careful not to generalize too broadly from the U.S. experience to Japan. Japan has borrowed from other societies for many centuries while maintaining its own unique and dynamic culture. In addition, one can point to differences as well as similarities in current trends in the two nations. For example, as Figure 2.1 indicates, Japanese women are much more likely than their U.S. counterparts to leave the labor force while they bear and rear their children. During this time the role of housewife assumes great importance. This includes an emphasis on careful meal preparation and the aesthetic

7

FIGURE 2.1. Labor Force Participation Rate for Women in Japan and U.S.: 1984

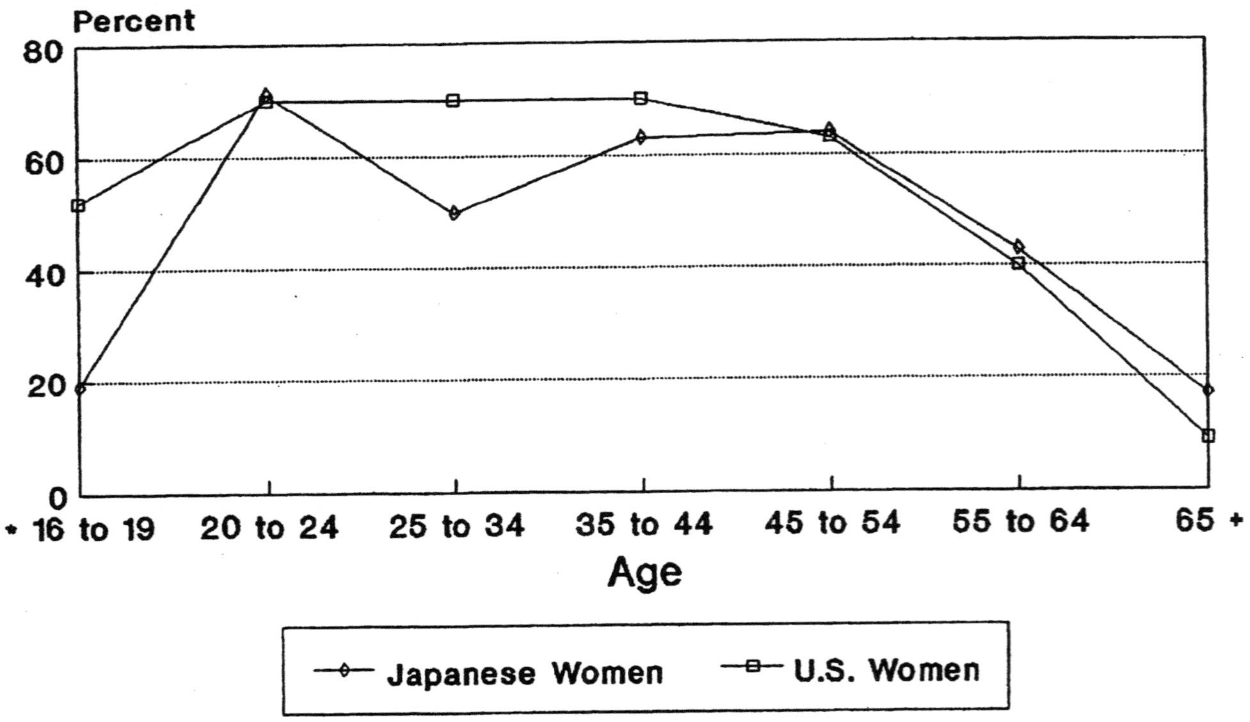

Percent

Age

— Japanese Women   — U.S. Women

* The category for Japanese women includes those 15 to 19 years old.

presentation of food. Many housewives shop daily, preferring fresh meat, fish, vegetables, and fruits (Vogel, 1978).

In addition, Japanese society is aging more rapidly than the U.S. Japan's birth rate is one of the lowest in the world–on average, Japanese women can expect to bear just 1.4 children, well below the 2.1 children considered necessary for replacement-level fertility (Population Reference Bureau, 1990). As a result the number of teenagers and young adults (those most likely to readily adopt different consumption patterns) is declining, while the number of persons over 40 is growing significantly (Cook, 1990).

FOOD SPENDING PATTERNS IN JAPAN

It is the difference in Japanese food spending patterns that caution against overgeneralization about the Westernization of Japan, particularly in terms of food consumption. Figure 2.2 compares the percent of the food budget spent on specific items in the U.S. in 1985 and in Japan in 1984. This shows that, as of the mid-1980s, Japanese households spent substantially more of their food budget on fish, vegetables, and cereals and grains (primarily rice) than households in the United States. On the other hand, U.S. households spent more on meat, dairy products, and eating out. An interesting difference between the two countries is highlighted in spending on fruits and vegetables. In the U.S., households spent an approximately equal proportion of the food budget on fruits and vegetables, while Japanese households spent more than twice as much on vegetables as on fruits.

Although these differences existed as of the mid-1980s, perhaps food spending patterns are changing in a manner that suggests rapid convergence with those in the U.S. in the near future. Figure 2.3 addresses the issue of the rate and amount of change in the percent of the food budget spent on major[1] categories of food between 1963 and 1984. As predicted by Bennett's Law, the percent of the food

---

1. Figure 2.2 omits expenditures on beverages, bakery goods, and oils and condiments. Each represented less than 10 percent of the food budget and has shown little change over time. Rice is a subcategory of cereals, but is presented because of its importance in the Japanese diet.

FIGURE 2.2. Percent of the Household Food Budget Spent on Types of Food: U.S., 1985 and Japan, 1984

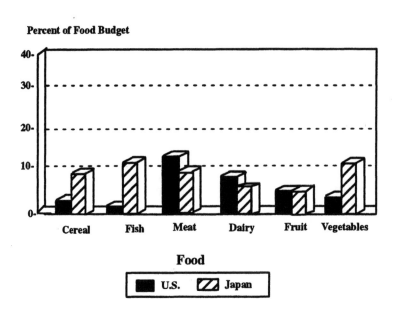

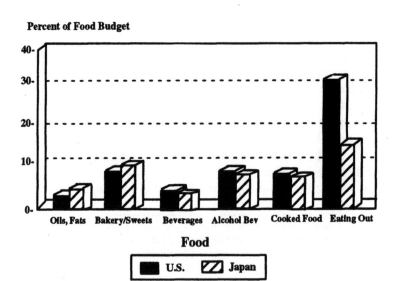

budget spent on rice (and thus on cereal products in general) has dropped dramatically since 1963, but most of this decline occurred before 1973 and has levelled off since that time. As persons familiar with Japanese eating habits will attest, rice remains a staple that is consumed at almost every meal. In 1984 Japanese households spent more of their food budget on rice than on beef, pork, and chicken combined.[2]

The percent of the food budget spent on meat and fish and shellfish increased between 1963 and 1984, but again these increases levelled off in the 1970s, while the percent of the food budget declined over time. Spending on fruits and vegetables has fluctuated over time, but spending on vegetables has consistently been about twice as high as spending on fruits.

Although not shown in Figure 2.3, a sizeable proportion of the expenditures in the fish and shellfish and vegetables categories in Japan is on products that are almost unknown in the typical American household. Approximately half of the expenditures in the fish and shellfish category are on fish paste and processed and salted fish. This includes several varieties of salted sardines; dried mackerel; and baked, fried, or steamed fish-paste cakes, bars, and patties. In the vegetable category, 40 percent of the expenditures (over 4 percent of the total food budget) is on dried vegetables and seaweeds, soybean products (primarily bean curd), and other processed vegetables and seaweeds. The percent of the food budget spent on these uniquely Japanese food stuffs has not changed substantially since 1963 (Cook, 1988).

The biggest change in spending patterns–other than the dramatic decline in the percent of the food budget going to rice–is in the percent spent on eating out and cooked food. However, this should not be construed as a Westernization of food buying patterns. Only 0.5 percent of the total food budget was spent on eating Western-style meals in 1984. Eating-out expenditures were concentrated on such items as noodle and sushi dishes and Japanese and Chinese meals. Consequently, in spite of changes in Japanese spending patterns over the last 20 years, they are still very different from the

2. The Total meat category in Figure 2.2 includes these meats as well as whale and other fresh meats and meat products, including ham, sausages, bacon, and canned meats.

FIGURE 2.3. Changes in Percent of Japanese Household Food Budget Spent on Major Categories of Food, 1963-1984

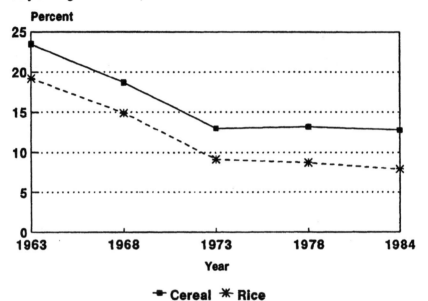

➡ Cereal ✳ Rice

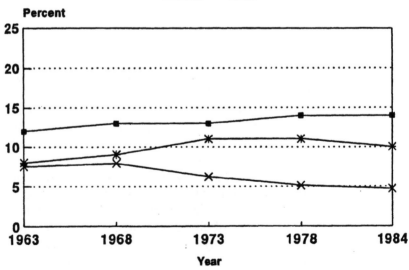

✳ Total Meat ➡ Total Fish & Shell ✶ Dairy Products

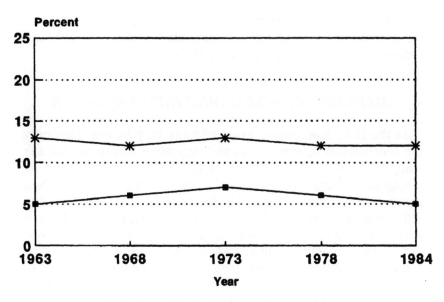

✳ Total Vegetables ▪ Total Fruit

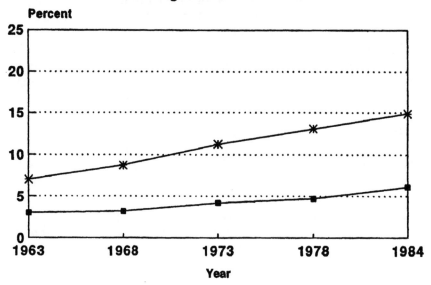

▪ Cooked Food ✳ Eating Out

Expenditures on beverages, bakery goods, oils, and condiments are omitted.

U.S. Although not presented here, a similar analysis was done on the quantity of food purchased by households. The results showed generally comparable trends (Cook, 1988).

DIFFERENCES IN SPENDING PATTERNS BY AGE

In the U.S., food consumption patterns vary by age. Generally, the consumption of fresh fruits and vegetables increases with age, while the consumption of beef, pork, and chicken peaks in the young-adult years and declines thereafter (Salathe, 1979). These patterns may or may not apply in Japan. It is important to consider differences in consumption patterns by age for two reasons. First, as noted earlier, some age groups are growing more rapidly than others. Second, producers may wish to target specific products for specific age groups. Available data for Japan are for households by broad categories of age of household head, rather than individuals.

Figures 2.4 and 2.5 show changes in the percent of the food budget spent on major categories[3] of food by four broad age categories between 1979 and 1984, illustrating some interesting variations by age and over time. Spending on cereals increased from 8 percent of the food budget for households headed by persons under 29 to 11 percent for those headed by persons 60 and older. This may be an indicator that older households maintain more traditional consumption patterns than younger households. However, spending on cereals decreased slightly for these oldest households between 1979 and 1984, while it remained unchanged for younger households.

Spending on meat did not vary consistently by age of household head, but spending on fish and shellfish generally increased with age. However, the proportion of the food budget spent in this area decreased for all households between 1979 and 1984. Spending on dairy products generally decreased with age, probably because younger households are the ones most likely to have small children. The proportion of the food budget spent on vegetables was higher in

---

3. As in Figure 2.3, spending on bakery goods and sweets, oils and condiments, and beverages is omitted. A more detailed analysis shows that spending on bakery goods and sweets declines slightly by age of household head. In the other categories there is very little variation in spending patterns.

the two oldest age categories and increased slightly for these groups between 1979 and 1984. Spending on fruit also increased slightly by age of household head.

The proportion of the food budget spent on eating out showed the greatest relationship to age of household head. It was highest for the youngest age group and declined considerably as household heads aged. In addition, spending on eating out increased between 1979 and 1984 for the two younger age groups, it remained constant for household heads aged 45 to 49, and it decreased for those 60 and older. This poses the question of whether the propensity to eat out will decline noticeably as Japanese society continues to age. Finally, spending on cooked foods showed very little variation by age of household head, and significant increases occurred for each age group between 1979 and 1984.

CONCLUSION

In the past, businesses marketing their products in Japan could depend on a rapidly growing population to increase their sales, but because of the extremely low birthrate, this is no longer the case. It is predicted that Japan's population will grow by just 4 percent in the 1990s, compared with 13 percent between 1970 and 1980. While novel products from the United States will continue to have appeal in Japan, some U.S. businesses may decide to target the daily purchases of the average Japanese household. To do this, they will have to acknowledge and market to the more traditional tastes of these consumers.

FIGURE 2.4. Percent of Japanese Household Food Budget Spent on Types of Food by Age: 1979 and 1984

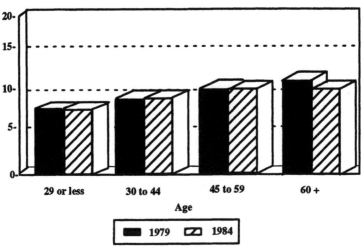

Cereals

Percent of Food Budget

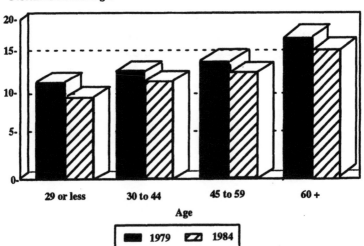

Fish

Percent of Food Budget

Meat

Percent of Food Budget

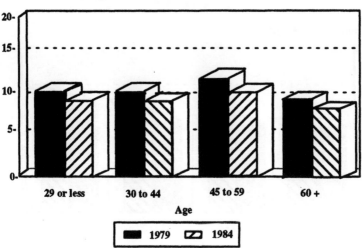

Dairy Products

Percent of Food Budget

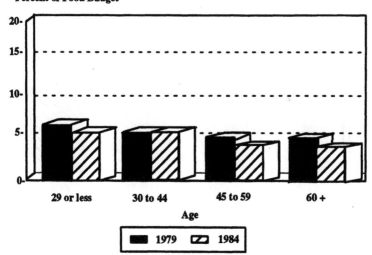

FIGURE 2.5. Percent of Japanese Household Food Budget Spent on Types of Food by Age: 1979 and 1984

Vegetables

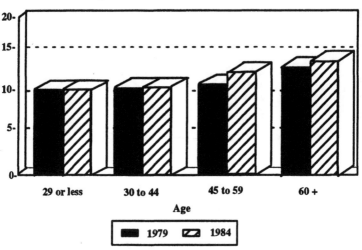

Eating Out

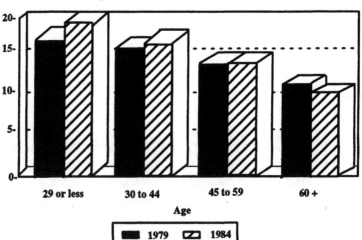

Fruit

Percent of Food Budget

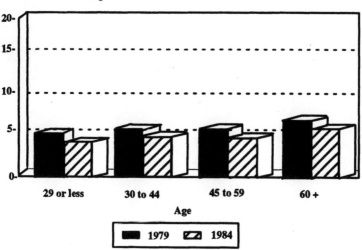

Cooked Food

Percent of Food Budget

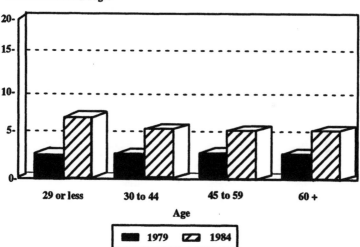

Chapter 3

Decision Making in Japan

Dorothy Z. Price

Although much of the work we previously have done has focused on the decision-making process used by final consuming units–individuals, families, and households–most of the formal research and less formal qualitative data provides information relevant to decision making at all levels. Actually, decision theory states that much of what is known about decision making indicates that the concepts are the same in any decision situation, whether it be for one individual, between or among individuals, within a family or household, within a business, between employee and employer, between business executives, or between governments; only some aspects of application differ. Therefore, in this paper, I will concentrate primarily on general patterns of decision making within Japan, in all aspects. Particular attention will be given to those levels above the level of final consuming unit, but which ultimately have a major impact on the final consumer. These include all aspects of business and government decision making.

APPROACHES TO DECISION MAKING

Three different general patterns of decision making are frequently encountered:

Traditional power approach. This is primarily a *win-lose* situation. One party in the negotiation holds greater power *or* is perceived to hold greater power. Therefore, that party gains as much as possible in the decision-making situation, while the other party gains nothing, or loses. It is important to recognize that perceived power is as important (and often, more important) than actual power.

Compromise. Despite the fact that compromise, as an approach to solving problems, has been seen by many as a positive force, essentially it represents a *lose-lose* approach. This is actually a process in which no one wins. Each party gives up something; in so doing, they arrive at a situation in which neither party is a total loser, but also neither party is enthusiastic about the agreed-upon alternative. Therefore, this often results in a situation where action seldom follows. In simple terms, no one feels sufficient motivation to act on the decision (because of what has been given up) and *nothing* is done.

Consensus. This represents a high-level, *win-win* approach to decision making. All parties involved in a problem situation gain something. Frequently, the solution reached represents a new and different alternative that was not previously considered.

DECISION MAKING IN JAPAN

Decision making by consensus is generally the norm followed in Japan. This contrasts with the U.S., where the prevalent mode is a traditional power approach, and a compromise approach is seen as a step toward a more inclusive or participatory style of decision making. A consensus approach is sometimes discussed but infrequently used. Because of this, Americans often find it difficult to understand the Japanese decision-making style. We still tend to be more accustomed to a top-to-bottom form of leadership.

In looking at this type of Japanese decision making, it is interesting to note that since it does not follow the usual American top-to-bottom style, we often tend to assume that it must be a bottom-to-top form of decision making. Actually, what exists is a *ringi* system, which involves circulation of documents and other information to a large number of people. All of these people, then, may put their seal on each document. The seal does not indicate that they approve, but that they have seen the information and do not actively oppose it.

The actual decision is made by extensive consultation; this involves wide, informal discussions with personnel at all levels. Everyone who may be involved with the effects of the decision in any way becomes familiar with the problem or the issue at this point, before the decision is made. All persons involved must have input and agree before an alternative is selected. This takes considerable time, often much longer than Americans are used to. Decision-making theory tells us, and the Japanese examples support the view, that this early involvement leads to a much greater willingness of everyone to participate in implementing a decision once it is made, and that implementation rapidly follows the making of a decision. This too contrasts with decision making in the U.S., where decisions are often rapidly made but slowly (and sometimes never) implemented. Bringing people into a decision-making process late or even after a decision has been made–or not allowing any *real* input during the decision-making process–frequently slows down implementation and may often totally stop it. Therefore, a decision may be made, but no substantive action follows.

The Japanese will not hesitate to postpone decisions if consensus cannot be reached. Postponement does not bother them, especially if the decisions are seen as minor. It is only with very important decisions, where major risks are involved, that the Japanese will be sure to decide and act in order to avoid the most negative possible results.

Since communication, both verbal and non-verbal, plays a key role in decision making, differences between Americans and Japanese need to be considered. Much data is available about communication in Japan. Important elements related to decision making include the emphasis on a definite structure of who communicates with whom, and how and why. The importance of harmony is also stressed. This is a key value in Japan and is sought by a subtle process of mutual understanding. Part of this is achieved through vague, ambiguous communication, which often is irritating to Americans. It, however, is not bothersome to Japanese, because they view this approach as a way of allowing different interpretations of what is said. These differing interpretations may avoid potential conflicts and, since avoidance of confrontation is important, they are seen as positive. The Japanese call this process *hara-*

gei ("art of the belly"), which centers on intuition and a feeling of what is correct and proper rather than on confrontation.

GENERAL POINTS RELEVANT TO DECISION MAKING WITH THE JAPANESE

Once one understands the basic framework of Japanese decision making, it can be useful to consider general suggestions that may be helpful in actually working with the Japanese in these situations. These include:

Japan is a vertical society. This is not a unique system, but it has some special aspects in Japan. There is almost always one single, distinctive relationship between individuals and between groups. The character of this relationship then underlies the etiquette and ethics of most behavior. In applied terms, this usually means that it is essential to try to obtain an introduction at as high a level as possible; but, one must also remember that the Japanese like to deal with outsiders of equal rank. Good advice is to use all available resources to determine the ranking of the person with whom one is dealing and to use patience and diplomacy to reach the highest-ranking person possible.

Business card exchange is a ritual. While business cards are important internationally, they are exceptionally so in Japan. Information must be correct and, preferably, should be presented in both languages. The business card exchange ritual also necessitates awareness of accompanying aspects, such as approaches to hand-shaking, eye contact, etc.

Research Japan, including its history. Many marketers are aware of the need to research product modification, distribution channels, and other elements that are necessary for entry into the Japanese markets. However, fewer take the time to research the culture and understand how its uniqueness affects transactions. In understanding Japanese culture, one can better realize why certain events occur (such as informal beginnings of conversations, the tendency of senior Japanese to remain relatively quiet, frequent side conversations in Japanese, much smiling) and also anticipate what other behaviors may occur. The cultural understanding provides a much more effective approach to being able to work with the Japanese than does simply reading a number of specific "tips on doing business."

Style is as important as substance. The way things are done is as important as what is done. One can imprint this in the mind by remembering elements of the Japanese culture, such as the tea ceremony and the presentation of food.

Harmony is all-important. The term *wa* is one of the most important words in the Japanese vocabulary and it means "peace and harmony." It pervades all relationships, and anything that breaks this harmony is seen as wrong. Because of this, etiquette is seen as a virtue and embarrassment as a sin. This view frequently results in a clash between cultures. Americans consider forthrightness and thrashing out opposing views as a virtue, while the Japanese have learned to speak in vague, general terms.

Decision makers require patience. Americans tend to think in terms of days, weeks, or months, while the Japanese think in terms of years and decades. Americans are proud of quick, decisive actions, while the Japanese are more likely to be deliberative and cautious. While schedules in the U.S. are seen as sacred, the Japanese tend to resist pressures on deadlines or delivery dates. When they do reach an agreement on dates, it is more likely to indicate the importance of harmony than to really represent a firm agreement in relation to the date.

Be aware of the importance of gift giving. Because of the gracious nature of the Japanese, gift giving is perhaps more important than in any other culture. Gift giving is a skill and an art, and here too, style is generally more important than substance. It is not the monetary value of the gift, but the presentation that is valued.

Recognize the role of entertainment. Entertainment will almost always occur in the evenings and can sometimes be quite lavish. These are not seen as times to conduct business, but primarily as times to bond friendships.

Generally, women will have lower status. Women usually have a difficult time achieving equal status in Japan, although advances have been made in recent years. Women who are in power positions in other countries will be accepted, but social situations may be strained.

Understand what is meant by yes. The Japanese word for yes is *hai*. But it does not necessarily mean assent or agreement. It means: "Yes, I hear you and I comprehend what you are saying. I do not necessarily agree with you." To understand this, one merely needs to remember the concept of harmony. A straight no is difficult for a Japanese to say since it disrupts the harmony of any situation.

Axtell (1989) actually lists ten ways the Japanese use to say no while still avoiding the word:

1. In conversation, a flat no will rarely be heard. On written forms where no one can be offended, it might appear.
2. When refusing more food or a drink, they will more likely reply in a positive form ("I'm fine") rather than "No, thanks."
3. Rather than say no, the Japanese will deflect the question with silence, make a sucking noise of air through their teeth, or perhaps utter a noncommittal "Oh, that would be very difficult."
4. Another device is the counter question: "Why do you ask?"
5. Appearing to ignore the question by shifting to another topic is a clear signal not to press further.
6. Some Japanese will equivocate simply to save face for the Westerner, assuming that he will understand the truth.
7. The qualified no might come in the form of "I will do my best, but if I cannot, I hope you will understand."
8. A favorite is the "Yes, but . . ." answer. One agrees, but then expresses his real state of mind; this is commonly recognized as no.
9. Delaying an answer, or "killing with silence," is another accepted method of conveying no. On the other hand, be patient–what you perceive as delay may just be the normal time required to arrive at a decision.
10. Among themselves, the Japanese may accept a task they know they cannot perform by saying "I accept," but add an apology that signals no. Apologetic words are usually negative verb forms. For example, the Japanese phrases for "I am sorry" are expressed literally as "I'm not able to assume that" or "It cannot be helped."

SPECIFIC HINTS FOR EFFECTIVE DECISION MAKING WITH THE JAPANESE

Before considering specific clues for actual negotiation or decision making with the Japanese, it is important to understand a concept made familiar by Edward T. Hall and Mildred Reed Hall (1987), who have done considerable study of communication. They distinguish between high-context and low-context communication. In high-context communication, information is primarily internalized or implied. In low-context communication, most information is in an explicit code. American culture is low-context; Japanese culture is high-context.

This difference can lead to many potential misunderstandings in negotiation. Americans search for the meaning of what is said and want to emphasize accuracy and specificity. Japanese are influenced by what is not said and send messages that Americans consider to be vague.

If this framework is remembered, a number of specific hints may be of use in decision making at any level with the Japanese:

1. Do not put a Japanese in a position where he can be seen as embarrassed or helpless.
2. Be modest and humble about what you, or your group, have to offer.
3. Always understand the concept of the need to save face.
4. Respect seniority and the elderly.
5. Expect periods of silence and/or little direct eye contact.
6. Remember that "proper" behavior is essential.
7. Avoid rigid schedules.

Negotiating and making decisions with the Japanese is difficult and challenging, but not impossible. Three key words should be remembered: patience, perseverance, and politeness.

Chapter 4

The Japanese Food Distribution System

Raymond A. Jussaume, Jr.

STRUCTURE OF THE FOOD DISTRIBUTION SYSTEM

The structure of the food distribution system in Japan is complex and difficult to analyze. This is due to the fact that the overall marketing structure, from farm to table, varies tremendously by commodity, region of production, and whether or not the commodity is processed. It is subsequently infeasible to make any useful generalizations about the Japanese food distribution system.

For example, the Japanese distribution system for processed foods is unlike that for fresh products because of the predominant role that large private firms play in the former. At the risk of oversimplification, it can be claimed that the distribution system for processed foods more closely resembles that for non-food Japanese consumer products than it does the system for fresh produce. It is apparent that *keiretsu* (an integrated hierarchical structure of firms) exists for a number of processed foods. One example is the distribution system for juice products. During an on-the-street survey conducted in November 1989 in Japan, some two dozen brands of canned juice products were identified. Those sold under the Suntory, Kirin, and Asahi labels were invariably sold through small liquor stores that sold, under an exclusive marketing arrangement, the alcoholic beverages of these three firms. Products sold under the Del Monte, Kagome, and Morinaga labels were customarily sold in small mom-and-pop variety stores. In other words, large firms in Japan that produce products for retail sale often have their own exclusive distribution system, and this pattern can be observed in some processed food products.

The source of the commodity in question also plays a major role in determining the route it takes from farm to table. In the case of apples, most of the fresh crop produced in the Nagano Prefecture is sold in season through agricultural cooperatives, because the Nagano crop is harvested before those produced in northern Japan and because of Nagano's geographical proximity to major urban areas. However, a large percentage of the apples produced in the Aomori Prefecture, which is located on the northern tip of Honshu island, are sold to private middlemen, who warehouse these apples for sale later in the year.

Distribution systems also differ by the particular commodity. Undoubtedly, the most distinct food distribution system in Japan is that for rice. The Japanese government continues to play a *direct* role in the marketing of rice in Japan. While some rice is sold through private market channels, the bulk of rice produced in Japan is purchased by the government, through the agricultural cooperatives. The rice is then sold to distributors at a price less than what the government paid for it. The history of this government involvement in rice markets dates back to the passage of the Rice Law of 1921 and the Rice Control Law of 1923, which were passed in response to severe shortages of rice and other food that sparked the rice riots of 1918 (Matsumoto, 1959).

The above examples are intended to demonstrate that there is no such thing as *the* food distribution system in Japan. There are a large number of food delivery systems and there are few commonalities that they all share. Therefore, in this brief presentation, my objective is to present a brief overview of the principal routes for the distribution of fresh fruits and vegetables in Japan. Currently, suppliers in Washington State ship asparagus and cherries to Japan, and there is hope that other fresh produce, such as apples, will be exported in the near future. Thus, it is important to understand how the domestic marketing system for fresh produce works in Japan. In addition, an appreciation of the Japanese distribution system for fresh fruits and vegetables can be utilized as an empirical foundation upon which to contrast future studies of the distribution system for other commodities.

OUTLINE OF THE FRESH PRODUCE MARKETING SYSTEM

The distribution system for fresh fruits and vegetables in Japan is characterized by a long history, geographical proximity between production and consumption areas, and a large number of producers, middlemen, and retailers. The focal points of this system are the wholesale auction markets.[1] Over 80 percent of domestic fresh produce (Table 4.1), and perhaps about 40 percent of imported fresh produce, are distributed through the 2,719 wholesale markets that exist throughout Japan. These markets are governed under the Wholesale Market Law, which was enacted by the Japanese Diet in 1971. This law replaced the Central Wholesale Market Law of 1923, whose passage can also be linked to the public dissatisfaction over the excessive control that middlemen and landlords had over food distribution and that was expressed in the nationwide rice riots of 1918.

Individual farmers are able, either directly or through local agricultural cooperatives, to sell produce at any wholesale auction market in Japan. In general, "a large proportion of vegetables are shipped through co-operatives when long-distance marketing (to marketplaces in major urban areas) is involved. Shipping by individuals is almost entirely limited to small, nearby markets" (Tanaka, 1989). The advantage of shipping through cooperatives is the higher prices paid by wholesalers and retailers in urban markets. By transporting their produce directly to local markets, farmers can avoid paying the transportation, boxing, grading, and mark-up costs associated with shipping produce through a cooperative. Farmers must therefore weigh their knowledge of various market prices, as well as the quality and size grades of their product, when making their decisions on where to market their produce. Most farmers use both strategies each season.

The existence of a large number of auction markets gives the Japanese farmer a number of alternatives for marketing produce. The same is true for retailers and wholesalers. For example, partici-

---

1. This report is based in large part on the more detailed analysis of Japanese Wholesale Auction Markets presented in Jussaume et al., 1989.

TABLE 4.1. Percentage of Domestically Produced Fresh Fruits and Vegetables Marketed Through Japanese Wholesale Auction Markets

VEGETABLES

Year	Percent Sold Through Wholesale Markets
1975	87.9%
1976	87.7
1977	87.4
1978	87.7
1979	88.9
1980	86.6
1981	86.8
1982	87.4
1983	89.0

FRUITS

Year	Percent Sold Through Wholesale Markets
1975	88.1%
1976	92.2
1977	87.5
1978	89.6
1979	88.0
1980	87.4
1981	82.7
1982	83.6
1983	84.6

SOURCE: Yamamoto, H. 1986. *Fresh Food Marketing*. Tokyo: Taiseichuo.

pating as buyers in the daily auctions in the new Ota market in Tokyo are 208 wholesalers and 2,213 authorized retail buyers. Each wholesaler and retailer concentrates on purchasing the quality and size grades of the particular commodities in which its firm specializes. Even large supermarket chains have discovered in recent years, after experimenting with direct purchases from farmers, that purchasing produce through auction market wholesale buyers is often preferable to making their own purchases. Supermarkets have learned that they can avoid the costs of training their own buyers, constructing their own warehouses, and purchasing size and quality grades of products that they do not want by working through auction market wholesalers.

The advantages to both producers and retailers of operating through wholesale auction markets are particularly pronounced in those commodities where freshness is a key component of product quality. There is a greater advantage in marketing fresh cabbages, lettuce, and cucumbers through auction markets than there is in potatoes or radishes. The nature of the auction markets promotes rapid delivery and easily discriminates between produce shipments according to freshness and other quality characteristics. This is because it is easier to receive a premium for superior produce. Commodities like potatoes, where freshness is easier to maintain and quality consistency is relatively easy, are more amenable to direct sales. Thus, the percentage of fresh potatoes sold in Japan that were marketed through auction markets in 1987 was only 29.6 percent, while the figures for cabbages and lettuce were 111.1 percent and 115.6 percent, respectively[2] (Ministry of Agriculture, Forestry, and Fisheries, 1987a, 1987b).

Imported commodities are sold through wholesale markets, although the percentage is not as high as it is for domestic commodities. For example, about 50 percent of imported sweet cherries were distributed through wholesale auction markets in 1988, although it had been about 60 percent in the early 1980s (Ministry of Agriculture, Forestry, and Fisheries, 1989). However, imported commodi-

---

2. These figures are greater than 100 percent due to underestimation of the volume of produce forwarded from Central to Local or Regional Wholesale Markets.

ties are generally not sold at auction. Frequently, sales are made directly between importing firms and market wholesalers, many of whom specialize in imported fresh produce. The frequent reason given by market participants for a lack of an auction in imported products is that the price has already been determined. Exporting firms are understandably nervous about consigning produce to an auction house for an auction sale.

RECENT CHANGES IN FRESH PRODUCE DISTRIBUTION

While there has been a slight decrease over the past ten years in the percentage of fresh produce sold through auction markets, more important changes appear to be occurring within the existing structure of the auction market-based distribution system than in the structure of the system itself. While there are still many small mom-and-pop type fruit and vegetable stores in Japan, the market share of supermarkets is increasing. Similarly, the percentage of produce sold at the largest Central Wholesale Auction Markets in the major cities is increasing at the expense of the smaller Regional and Local Wholesale Markets. This is true not only in absolute volume, but particularly as a percentage of the highest-quality merchandise. For example, it is not uncommon for a wholesaler at the Setagaya market in Tokyo (a Local Wholesale Market) to purchase produce from a wholesaler at the Ota market, which, as the largest wholesale market in Japan, has a tendency to attract the best-quality produce. This means that auction houses and wholesalers at the biggest Central Wholesale Markets are developing a dominant position vis-à-vis participants in other markets around the country.

One indication of the enduring strength of the auction market system, as well as the increasing concentration within it, is the opening in April 1989 of the new Ota Central Wholesale Market in Tokyo. The Ota market has 386,000 square meters of floor space. This is more than ten times as large as the Kanda market, which it was designed to replace. Daily sales of fresh produce are approximately 3,000 tons, as opposed to an average of 2,000 tons daily for the Kanda market in 1987. Since the Ota market opened, competition among producers is said to be more intense than it was in Kanda, because more varieties of produce are being gathered into the marketplace. Supermarkets are pleased with the Ota market because they can gather large shipments of a wide variety of commodities in one place. However, although there is no empirical evidence to corroborate this assertion yet, it may be that the market positions of the other Wholesale Auction Markets in the Tokyo metropolitan area have been weakened.

Another indication of how the internal structure of the Wholesale Auction Market system is changing is the growth of the *Sakidori* method of making sales. *Sakidori* (literally, "taking before hand") is a form of buying that was implemented for the benefit of supermarkets and wholesalers in distant city markets, so that these buyers could purchase and transport produce from the marketplace to their stores in time for their morning business hours. The *Sakidori* system also benefits producers and shippers who find it easier to send entire shipments to a single marketplace, like the Ota market, instead of dividing up the shipment and sending it to different local wholesale markets in a city.

In *Sakidori*, wholesalers make a promise with an auction house to buy a section of a particular quality and size lot(s) from a shipment of produce from a producing region. The price is set at the highest price received for produce from that lot at that morning's auction. *Sakidori* is restricted to 30 percent of any given lot, although market participants informally admit that the actual percentage is higher, perhaps as much as 50 percent. The rationale for the restrictions on *Sakidori* is that if there were no limit, small retailers would find themselves excluded from access to particular lots. However, these restrictions are being flaunted because supermarkets are becoming very important customers for wholesale agent companies. In this manner, supermarkets, and the wholesale firms they do business with, are becoming the dominant players in the wholesale markets. In addition, as the premium produce of a given lot is often taken out by *Sakidori*, the prices set at auction may be somewhat lower than they would be otherwise. This means that supermarkets can often buy high-grade goods at a comparatively low price and that small retailers have limited access to the best produce, which further weakens their ability to compete for customers.

THE SANCHOKU *MOVEMENT*

In addition to changes occurring within the Japanese food distribution system, there are some being made without. Perhaps one of the more interesting has been the growth of the *Sanchoku* movement. This term actually describes a wide variety of organizational forms, with one interesting similarity: the *direct* sale of produce from farmers to consumers. This strategy is utilized quite frequently, although not exclusively, for the distribution of organic food products. By maintaining a personal tie with specific producers, consumers can be assured that the produce they receive is grown in exactly the way they want. Producers receive the benefits of the elimination of middlemen and premium prices for their produce that they would not get otherwise. Since these transactions occur outside the jurisdiction of governmentally supervised markets, there is no accurate data available on the extent of the growth of the *Sanchoku* movement, but it is thought to be significant (Tanaka, 1989).

The growth of the *Sanchoku* movement is an indication that one of the biggest forces promoting change in the Japanese food distribution system in the future will be changes in the pattern of consumer demand. As Washiyama (1989) has pointed out, Japanese consumers are demanding even higher quality and variety as well as less use of agricultural chemicals in the produce available at retail stores. Consumers are also demanding more convenience and pre-preparation of foods. As producers, wholesalers, and retailers strive to meet these growing consumer expectations, new market lines and corresponding market delivery mechanisms, such as the *Sanchoku* system, will be developed. One of the challenges for the future for overseas producers and shippers will be to keep abreast of these changes and adapt to them.

CONCLUSION

The objectives of this brief report have been twofold. The first was to point out that there are many types of structures utilized in the distribution of food in Japan. The second was to present a brief outline of the dominant system used in the marketing of fresh produce, the Wholesale Auction Markets. While this system is in constant evolution, the auction markets continue to be viable in Japan. In other words, not only can Japanese patterns of consumer behavior modernize without becoming Westernized (Jussaume and Cook, 1989), but so too can the Japanese system for delivering agricultural commodities to consumers. While the number of participants in the system will continue to decrease, and the process by which sales are made will be altered, it is reasonable to expect that the Japanese food distribution system will maintain distinct features that overseas producers and sellers will need to adapt to in the future.

Chapter 5

The Role of a State Department of Agriculture

Eric Hurlburt

I'd like to thank you for being able to speak here. I am the new kid on the block in the Washington State Department of Agriculture, in the Marketing Division. I just came on in April, so I'm not any great expert. I come from the regulatory side of the fisheries business. My main claim to fame right now is that I just completed a year of living in Japan, which gives you a little different perspective than it does for people who do business over there by flying into Tokyo, suffering through Narita Airport, spending some time in hotels and restaurants, eating things that they never saw before, and then flying out. Tokyo may be the gravitational center, the black hole of Japan sucking everything into the center of the country. It is not the only market, and there is a lot of stuff going on in other parts of the country, which are different. I agree with Ray Jussaume that you see variations from one end of the country to another. It is a much more diverse market than a lot of people realize.

What I would like to do today is talk a little bit about what the Department of Agriculture is doing to try and aid the food industry of Washington State in selling their food products to Japan. We have been in the business of helping the agricultural and food industries in selling their products for some time. In 1985, we expanded and set up an office in Tokyo; Japan is obviously the major market for Washington State products. We've recently expanded again and are picking up seafood, which is my main interest. Forty-three percent of agricultural commodities going through Washington or Northwest ports goes to Japan. It's closer to 75 percent for seafood.

Japan is obviously a focal market, although I think we must keep in mind that it is not the only market for our products. One of the missions we have is to provide a diverse, stable, and profitable market for Washington's food products. We aren't looking just at Japan, but it is the key player.

I should also point out that there are other agencies involved in the Japanese market. They are assisting in the promotion of Washington products, as well as United States products in general. One of these is our state Department of Trade and Economic Development. They are co-located with us in Tokyo, and have an office over there. On staff they have Bryson Bailey, who is a refugee from the Department of Agriculture. He still thinks like an Aggie, although he works with the Department of Trade now. We have close working relationships with them, and with the limited resources both agencies have for doing this market work in Japan, each of us has a one-man office. We try to pool our efforts and work cooperatively together. Other major resources are the U.S. Department of Agriculture's Foreign Agriculture Service and the Agricultural Trade Officer (ATO) in Tokyo. They do everything from giving us a lease to making business introductions. They conduct trade shows, invite people over to Japan, and help arrange tours coming back this way. They are very active in promoting U.S. products, and they work very closely with our department and our efforts.

We have five program managers here in Washington, all located in Olympia. Each deals with different commodities. All of us deal to some extent with Japan. We have people specifically assigned with Japan as their main focus. I inherited some of that responsibility just by having been over there for the last year. Kent Nelson, who runs our Tokyo office, is the point man. Obviously, it is very difficult for one guy to cover all of Japan, which, as I said earlier, is not a homogeneous market. There is a great difference in the markets between Hokkaido (in the northern part of Honshu) and Kagoshima (in southern Kyushu).

And then Tokyo is just something unto itself. What we do is what any marketing and trade promotion organization does. More than anything, we are a funnel for information going back and forth from Japanese buyers and sellers to Washington buyers and sellers. Probably the bulk of my activity lately has just been getting inquiries that have come through Kent Nelson in Tokyo. Some company may want fishmeal, or be looking for black cod. Other times, companies are looking for asparagus or cherries or whatever. Especially when the supply is short, they are really looking hard. We try to line them up with prospective trading partners.

We spend a tremendous amount of time with companies on both sides of the Pacific, trying to advise them how to do business with one another. Japan is a formidable market for many companies, not only for a small company that is trying to learn the export business (which is something in itself) but also for any company then having to deal with the Japanese culture and all the other difficulties of doing business over there. We spend a lot of time–especially with small and medium companies, and with people who are new to exporting–on how to do business, on developing possible contacts for them; trying to maintain databases on people who might be interested in dealing with them; advising them on labelling requirements and various rules and regulations they have to deal with; and on pointers to doing business. Sometimes it gets very basic, down to helping them decide whether they should even pursue getting into the Japanese market at that time.

We have some people who, as soon as they have made their first sale domestically, immediately start thinking, "Export, that's great." But, you have to have the resources to persist in that market for some time before you are really ready to jump into it. One of the biggest shortcomings we see is that people expect to go over there, do one trade show, and make a killing. Even the experienced people I know who go over there may do the trade shows and all the follow-up work, yet it will take them three years before they make a significant sale. That doesn't mean it's a profitable sale, it's just their first significant sale.

We also do the same sort of function in advising Japanese firms. There's a lot of interest on the part of Japanese firms in sourcing more product from the United States. There is a growing interest in more value-added products. As we heard earlier, the demographics of Japan are that there aren't many young people. I think the median age in the farming community is something like 60. It's the same thing in the fishing industry. The young guys can't find wives because the young women have all gone to the city for the fast life.

They're having some real problems in domestic production, especially in the labor-intensive activities.

More and more Japanese companies are looking at going offshore, which includes coming over here, to locate processing facilities and do the value-added work, and then ship the products into Japan in a form that can be set right on the supermarket shelf. We work with advising these companies on the opportunities over here. We work closely with the Department of Trade on possible investment activities and, again, try to match up buyers and sellers. Some of this also includes organizing tours for various Japanese individuals or groups of businesses that want to come over here. We recently had a group representing the major Japanese breweries over here looking at hops. Last week, I spent some time with Japanese fishermen from the Shizuoka Prefecture, who were over looking at what we're doing. We've recently been developing a rather close relationship with a group in the Miyagi Prefecture, up in the northern part of Japan. They are very interested in both timber and fish.

Another thing that Kent Nelson (our Tokyo representative) does is market analysis. Someone in Japan who goes out to lunch with the Japanese, who reads the Japanese newspapers, who sees Japanese TV, who wanders through the supermarkets and rubs shoulders with people on the street, has a much better feel for what's going on over there than people sitting over here and reading all the printed information that comes over. There's a time lag, and as was discussed in the first presentation, communication is different over there. What you see written down is often just what's on the surface. It doesn't explain what's going on down deeper. Getting a broader view, more of a gut feel for things, is really important. Having someone doing that on site is critical. Kent is constantly feeding information about what's going on to us and to companies he has direct contact with here in the States. He develops reports. He's recently done one on hops, one on Walla Walla sweet onions, and a third in which he identifies what agricultural products have particular opportunities in Japan because of seasonal factors. We then pass those reports on to our different industry clientele.

Another aspect of the work on both sides of the ocean is trade shows. Trade shows are important to doing business in Japan. Also, the trade shows can be very big, as anyone who has been to Foodex will know. There are trade shows going on all over the country. More and more areas are expanding or setting up their own trade shows, partly because it lures people to their part of the country–to a place like Fukuoka, down in the extreme southern part of Japan. Servicing all these trade shows is, of course, a problem. Travel in Japan is expensive and there are all sorts of hassles in terms of getting product moved around. There are also lots of opportunities. You must show up at those trade shows consistently, year after year. The first time, you get people just looking and nibbling on your samples. The second year, they'll talk to you and they may order a small shipment just to check you out. Then maybe the third year, they'll come back and, if you're still there, they start thinking about doing some real business. If you go there two years, or worse, one year, you've wasted your airfare. We help recruit people to work in those trade shows, both from the Japanese and the U.S. side. We also work with the ATO's office at the U.S. Embassy in setting up these fairs. Again, we work with the companies trying to maximize their possibilities of success.

One other area where we have a unique opportunity is in governmental relations. Government officials are viewed quite differently here and in Japan. Here, if I go out to a bar and run into someone and start having a conversation, and they ask, "What do you do?" and if I say, "Well, I work for the state," that's often the end of the conversation. In Japan, you say you work for the government, and that is a very honorable position. Even though it's not the best-paying position in the world, it tends to attract the cream of the crop out of the universities. It is something people aspire to. By being a government employee, you have a certain amount of respect and prestige, which allows you to get through doors that someone else might not be able to penetrate. We are in a position where, when companies want to come over and do business in Japan, we can help make introductions for them. In Japan, you can't just make a phone call and say you'd like to talk to someone in a Japanese company. You need someone to introduce you, to basically testify to the goodness of your character. I'm sure you all realize business over there is based more on personal relationships than on the bottom line.

We also are in a position to work with the Embassy, the U.S.

Department of Agriculture, the U.S. Department of Commerce, and other groups that are working to try and reduce trade barriers. Our director, Dr. Pettibone, has been in Japan several times trying to get beef quotas adjusted. We are seeing more and more products getting into Japan, more quotas being reduced, and opportunities increased. A lot of that has been supported by our involvement: getting in there and making the day-to-day contacts with these people, explaining our position, and constantly jumping through all the appropriate hoops that are presented repeatedly.

One other area we focus on is coordination between government agencies. There are a lot of groups from Washington State that are involved in the trade picture in Japan. Many of the IMPACT (International Marketing Program for Agricultural Commodities and Trade) Center researchers that have been over there have worked closely with Kent Nelson and with our Olympia staff as a source of information. This also allows us to help coordinate what is happening on the research side with what we know is happening on the business side and at the government level. We can act as a listening post to find out what is going on and then try to disseminate that information.

Part of our efforts are financed from state general funds. Indirectly, we also get a fair amount of money from the Targeted Export Assistance (TEA) Program. We've got four projects that have been going on recently. One deals with seafood, one is introducing our wines into Japan, one deals with asparagus, and another one with dairy products. Those have been going along very successfully. Of course, you are always subject to the variations in policy out of Washington, D.C.

The seafood program involved working with one of the restaurant chains in Japan. Money from the TEA fund paid for printing various promotional materials, menus, and that sort of thing, to the tune of about $30,000, which generated about $5 million in the sale of seafood–not strictly from Washington, but from the Pacific Northwest. With the dairy products, we've just done some market research. We'll be passing the resulting information on during the second year to try to help Washington producers in cheese, ice cream, and other dairy products get a leg up on getting into the Japanese market. I think Washington wines have a great potential over there, and we are gradually introducing them to more and more places. It's a growing segment as the country becomes more wealthy. It's a prestige item, and Washington has the wines to fill that niche.

One other important measure of success for us is sales resulting from our efforts. Keep in mind that we have to get renewed funding every other year. Last year, our Tokyo office generated or contributed to the equivalent of about $4 million in sales. Over the next three years, those same sales should grow to a total of about $20 million. Also, we've introduced companies who are now investing in Washington State. For a one-man office with some support activities out of Olympia, I think that's a hell of a track record.

I'd like to mention one other thing that's going on in Japan. I'll mention it partly because I, too, was drafted to be here. Desmond O'Rourke invited my boss, Art Scheunemann, to make this presentation, but Art is in Japan right now. We have a sister-state relationship with the Kyogo Prefecture, which is the prefecture incorporating the city of Kobe, just west of Osaka. This is another example of some market opportunities over there that are sometimes missed by all the focus on Tokyo. Kobe itself has about a million-and-a-half people. It is a very international city by Japanese standards. It has a great potential for the sale of Washington State products, because as a sister state, Washington is a known entity. Many of its people have travelled to Washington State, people in the government know us, there is a close relationship. It's a good place for small companies to get into. Art is over there right now, along with some people from the Department of Trade, talking about some of the possibilities for setting up activities that would help companies enter the Kobe market. Besides Kobe itself, within a thirty-mile radius you have the cities of Osaka and Kyoto, and a total metropolitan market of about ten million people.

To wrap up, I'd like to say something about some of the things we need to do to expand or improve what's going on in our marketing efforts. Japan is going to remain an important market for our products for some time. We need to keep track of what our overall goals are in marketing. Ultimately, for us as state employees, this is to promote the economic and social benefit of our citizens. We happen to do it by trying to promote sales, stabilize markets, and ensure that the agricultural industry is getting the best return for its products, a return that will be both stable and long-term. To do that, we need to have consistent marketing policies, programs, and funding.

I recently read a report by the Northwest Policy Center at the University of Washington that highlighted the importance of consistency. In Japan, where the government doesn't change very often, they can set up ten- and twenty-year programs, in which they can follow a policy for ten years and then do mid-course corrections. The policy is set and the funding is assured. We at the state and federal level live in a much more variable situation. We cannot commit very far ahead because we may not have the funding and resources needed to meet those commitments. In Japan, such variability makes it extremely difficult to operate. You need to have a consistent base there. You need to establish that personal rapport by having someone in our Tokyo office whom people get to know over a period of years. He has had doors open, which allows more doors to open. If you jump in and jump out, you'll never succeed.

Another critical item is the coordination between all the groups that are working over there. We in the Washington State Department of Agriculture have good communication between us and the Department of Trade, and with the IMPACT Center. We need to expand upon that coordination and that communication to assure that all the information that different people gather is somehow circulated and that we are all marching forward in an orderly manner, instead of just a bunch of people out "shotgunning" it. This means we need to spend more time working with our clientele group and finding out what the industry wants to do, as well as spending more time with people over there and finding out what the opportunities are.

Chapter 6

The Japanese Market for Beef Products

David Youmans

Agricultural exports from Washington State are valued at rough-
ly $1 billion per year. Of that total, beef cattle and its by-products
rank second only to wheat in total export value and vie for that
position from year to year with world-famous Washington apples.
Japan remains overwhelmingly the number one client for beef products, while Canada is a somewhat distant second. Thus, it is in this
context that the IMPACT (International Marketing Program for
Agricultural Commodities and Trade) Center–in its collaborative
effort with other agencies and with Cooperative Extension (through
its network of agents and specialists)–has worked with the entire
cattle industry toward gaining a greater share of the Japanese market for Washington beef.

BACKGROUND

In the years prior to 1988, American beef producers became all
too familiar with a climate of restrictive practices in the Japanese
marketplace that not only severely limited foreign access to that
market but also created and sustained unreasonably high retail
prices for Japanese consumers of beef products.

In that environment, the Livestock Industry Promotion Corporation (LIPC) effectively protected the high-cost local production of
beef by levying severe tariffs and surcharges on foreign products.
Those revenues were used to subsidize Japanese cattlemen and to
sustain artificially elevated prices for locally grown beef.

47

As many of us know, those Japanese cattle were of two primary types. The indigenous Wagyu–produced in confined quarters on a small scale and intensively finished on imported (mostly American) feed grains–provided 36 percent of locally slaughtered beef. Dairy beef, largely Holstein, accounted for the remaining 64 percent of the national kill. Both breeds were fed to excessive finish, thus forming the highly marbled meats preferred by Japanese consumers. Only the Wagyu attained the A grades by Japanese standards, while dairy beef differentially graded across the B band.

It is important to note that even on such a highly restricted and orchestrated market, American beef had successfully achieved an enviable position by late 1987. Of the three major exporters to Japan, the U.S. was second only to Australia and far ahead of third-place New Zealand. Since Australian beef at that time was grass-fed, America was the undisputed purveyor of grain-fed, high-quality beef and had truly consolidated its middle niche in the Japanese market. Significant, too, was the fact that Japan was, by a large margin, America's and certainly Washington State's number one export market for beef products. Some 80 percent of U.S. export beef went to that country.

THE LANDMARK EVENTS OF 1988

At the same time that it was participating in the Uruguay round of the General Agreement on Tariffs and Trade (GATT), the U.S. petitioned GATT to study illegal trade practices by Japan. The Japanese responded by negotiating a new U.S.-Japan Beef Agreement, which was signed on July 5, 1988. The chief American negotiator of this agreement was Dr. Clayton Yeutter, the U.S. Trade Representative at the time. Of continuing significance to the American beef industry was the Bush administration's appointment of Dr. Yeutter as Secretary of Agriculture.

Central to the new agreement was the phase out of the Japanese import quota system by 1991. The LIPC would likewise phase out its own role in the regulation of beef imports. As the quota system wound down, the quotas for imported beef would themselves increase until effective restrictions no longer existed. Japan, however, reserved an emergency adjustment measure that could be implemented if, in any year, beef imports exceeded 120 percent of the previous year's imports. That reservation expires in 1994, at which time Japan will presumably comply with the provisions of the Uruguay round of GATT.

While the 1988 agreement (which increased quotas by 60,000 metric tons yearly over the following three years) was a welcome turn of events for American beef producers, then Secretary of Agriculture Richard Lyng quickly pointed out that it was also good news for America's competitors in the beef trade. The absence of quotas opened the doors wider to all exporters of high-quality beef. However, in the judgment of Washington cattlemen, feeders, and packers at meetings in Blaine and Yakima, local producers seemed extremely well positioned in terms of quality, reliability, and proximity to benefit from the new conditions.

A second event in the State of Washington that followed the agreement by a matter of mere days was the acquisition of Washington Beef, Inc., by Farmland Trading Co., Ltd., of Tokyo. That transaction placed the state's second-largest meat-packing capability directly and wholly under Japanese ownership. And while the former feedyards associated with Washington Beef–namely Schaake, Monsons, and Van de Grafs–continued to feed independently for themselves and for the Iowa Beef Processors (IBP) at Wallula, there was activity on the part of Japanese principals toward ownership or partnership in Washington feeding operations. Washington Beef, Inc., itself began to undergo successive changes in its transformation toward Japanese business practices.

IBP at Wallula continued to be the state's undisputed leader in the beef-packing industry, killing some 2,000 animals a day, six days weekly, and fabricating an additional 1,200 to 1,500 carcasses from its Boise plant. L & M Feeders, Inc., also at Wallula, remained IBP's principal source of slaughter cattle. At that time IBP disclosed that while only 3 percent of their chilled boxed-beef was destined for the Japanese market, fully 11 percent of their total beef products were so shipped. That disclosure demonstrated the immense importance of beef cattle by-products entering that export market.

In September 1988 the U.S. Meat Export Federation staged its second annual "U.S. Meat Month in Japan." That event was representative of the proactive kind of promotion required to sustain and expand the American share of that growing market. Beef interests in Washington would be served well by following suit as such opportunities continue to arise.

THE DYNAMICS OF 1989

The year 1989 was the first of the new U.S.-Japan Beef Agreement. Perhaps as early as the end of March, signs and trends would indicate how America was faring and, in that context, how Washington was performing. Quality, reliability, and proximity all worked to Washington's advantage, but the industry could not be passive and expect to excel. It had to encourage feeders and packers to miss no opportunity to promote local beef products among the Japanese. And, where and whenever possible, ranchers themselves were compelled to do likewise.

Meanwhile, Japanese trading companies continued to seek overseas production bases–including ranches, feedyards, and packing plants. That type of activity was seen in Australia as well as in the U.S. As Japanese import quotas relaxed, overseas Japanese operations sought to compete with their own producers of dairy and Wagyu beef by offering well-marbled foreign beef. Some companies began to use Wagyu semen to inseminate Angus cows, and Washington State University's (WSU) Wagyu Beef Project scientists became involved in genetic research here at home. The competition was active, and the formerly well-protected Japanese beef industry faced internal challenges to its survival.

The much-hailed Beef Agreement, which had permitted entry of 60,000 metric tons per year of additional beef into Japan, began to augment the simultaneous buy-sell component of the market, and it set import duty levels for the three years after liberalization at 70, 60, and 50 percent of landed price, respectively. The LIPC continued to play an omnipresent role in market dynamics. Uncertainty continued to flourish as to what forces would influence a freer market in the early 1990s. The U.S. share of the Japanese beef market climbed from 29 percent in 1984 to 35 percent in 1986 and 48 percent in 1988, and it reached nearly 50 percent in 1989. (That significant gain was essentially at the expense of the Australian share–despite heavy Japanese investment, joint ventures, and increased grain feeding in Australia.) A number of factors contributed to these U.S. gains: (1) active and excellent promotional activity by the U.S. Meat Export Federation; (2) breed-specific campaigns, such as those for Certified Angus Beef; (3) company-specific activity, such as IBP's program for selling full carcasses in 22 meat cuts; (4) various other branded-beef efforts; and (5) Japanese acquisitions and resultant directives, as in the case of Washington Beef, Inc.

THE WASHINGTON STATE UNIVERSITY COOPERATIVE EXTENSION INTERVENTION

Given the above scenario, it was vital that selected Extension specialists and agents at WSU gain personal exposure and experience in the Japanese production system, market structure, distribution dynamics, consumption patterns, and consumer habits. That knowledge could then be translated into Extension education programs designed to bring Washington Extension agents, producers, feeders, and packers into a stronger position with respect to the realities of the Japanese marketplace. It was appropriate at a time when American access to the Japanese market was being enhanced, Japanese acquisitions in the Washington beef industry were taking place, cattle feeders were becoming involved, animal scientists were looking into feeder cattle genetics, and ranchers were eager to play some role in exporting. WSU established a Japan Beef Study team that spent field time in dialogue with feeders, packers, and beef-industry representatives in the state, exchanging ideas and inventorying the concerns of those involved. That input was incorporated into a study tour of Japan–which, in late 1990, was still contributing important findings to the industry.

The team undertook an intense study of the Japanese beef market in September and October 1989. An itinerary had been developed with the cooperation of the U.S. Meat Export Federation (USMEF) of Denver, Colorado, whose Tokyo office staff assisted with schedules and appointments. The WSU team leader had requested direct experience with the dynamics of Japanese beef production, feeding operations, carcass auctions, slaughter and fabrication, import marketing, wholesaling, merchandising, retailing, consumption patterns, and consumer behavior. He requested an opportunity to share both Oriental and Western cuisine containing beef products.

With USMEF's assistance this field program was executed in its entirety–with the added components of a live cattle auction, livestock show, TV interview, and exchange of views with university faculty members thrown in for good measure. More than 50 resource persons, organizations, and facilities were visited during the study. Some on-site impressions are capsulized below:

The initial perception in Japan is that of affluence, at least as evidenced in urban Tokyo. Money isn't being thrown around carelessly, and doubtless there are many Japanese who don't have as much spending power as they would like, given current prices of consumer goods. But, generally, there is an air of prosperity, along with responsible consumerism. Shoppers, both men and women, seem to choose carefully and expect value for money spent. Quality is everywhere, from expensive shops on the Ginza to neighborhood food markets. It is common to see Japanese Wagyu and dairy beef in the market for $25-30 per pound. Quality imported beef, to date largely American, satisfies a 33 percent middle niche in a market of rising expectations. Those high quality beef products are consumed partly in Japanese dishes, so that recognizable American-type cuts are soon lost in the cultural milieu. However, there is an important steak market, as well, in mainly Western-type hotels and fine restaurants. Yet a third substantial market for beef is in the fast-food industry. This segment has traditionally been supplied by Australia with grass-fed beef. In all cases, beef consumption in Japan is overwhelmingly by Japanese consumers, regardless of significant numbers of expatriates in the country. If there is a single perception which should send America and Washington State some unmistakable signals, it is that Australia is perfectly capable of producing well-marbled, grain-fed beef, is doing so, actively promotes it on the Japanese market, and fully intends to compete in that now less-restrictive-quality-beef import market. Furthermore, it is commonly understood that Japanese trading companies themselves are actively involved along that supply line. Optimally,

Washington producers can, and must, be competitive in that scenario, but complacency is certainly not the watchword of the day (Youmans, 1989).

RAMIFICATIONS FOR WASHINGTON STATE PRODUCERS

There appeared to be a number of implications for Washington beef producers as assessed through the consolidated findings of that WSU Japan Extension Beef Study Team. There had emerged over time a body of "conventional wisdom" about the Japanese beef market that seemed to vacillate between superlatives and generalities. The findings of the team tended to moderate the superlatives and bring variability to the generalities. There was a great deal of mythology about the Wagyu cattle industry that required a measure of realism in its retelling, and the variety of Japanese eating and buying habits certainly challenged some of the stereotypes.

There were clearly some valuable signals for a select few Washington producers of feeder cattle in all this. There were opportunities for a few feeders to do some innovative things. One or two packers could address food safety, shelf life, and niche-market specification components of that share of their kill destined for the Japanese market. The market is not for everyone, and its demands would require time, investments, and risk. The media described a current surplus of imported beef in Japan that was proving difficult to retail. Since the Japanese were obliged to honor increased import quotas before LIPC relinquished price and supply controls, that surplus was not surprising. It was, however, troublesome.

COOPERATIVE EXTENSION PROGRAMS

Some products of subsequent and current Extension export-enhancement activities bear mention. Along with Iowa State's collaborative work with Midwest Agribusiness Trade Research and Information Center (MATRIC) and some good work at Idaho and New Mexico State, the WSU Extension team published "The Japanese Beef Market: Implications for Washington Producers"

(Nelson et al., 1990), which is available as EB 1567 on the WSU campus.

Extension faculty likewise presented some seventeen timely public-education programs, including those at the Washington Cattlemen's Association, the Washington Cattle Feeders' Association, the Washington Beef Commission and Ag Expo/Farm Forum 1990, and the National Association of County Agricultural Agents.

THE WASHINGTON STATE BEEF INDUSTRY RESPONDS

Though it is an increasingly elusive exercise to chart credit for anything innovative in the livestock industry, the Cooperative Extension at WSU takes a great deal of pride in its role as an information and resource broker in the realm of export marketing of beef. Documentable practice changes have taken place at two Washington packing plants with respect to increasing the shelf life of beef products on the Japanese market. Such factors as temperature, film, purge, portion size, sub-primal handling, color and strength of boxing, and shipping efficiencies were addressed concurrently with the disclosure of Extension findings on those very issues.

Carcasses intended for shipment to Japan were split and graded according to Japanese standards and practices subsequent to Cooperative Extension education programs on those topics. Feedlot practices involving age, length of feeding period, rations, monitored rate of gain, meat quality assurance, slaughter weight, and genetics of feeder cattle were addressed in concert with Extension findings offshore.

To further describe Washington's response, Washington Beef, Inc., took steps to comply with Japanese specifications for primal and sub-primal fabrication; exchanged expertise to enhance market awareness; made full use of a Japanese distribution network; and made progress in the food-safety and shelf-life areas. As a result, their percentage kill for export, while dropping from 13 percent to 5 percent after acquisition, increased to 30 percent (50 percent seems to represent a realistic attainment if the company is so inclined). There were certainly active and creative ideas on boardroom tables in Tokyo in this connection. Associated with Washington Beef activity were the feeding contracts in progress at Monson Ranches, where careful cattle selection, specified rations, and extended feeding periods were all critically important. What part IBP Wallula is playing in the 22-cut full-set program is not immediately known, but their proximity to Japan's nearest deep-water ports certainly leads to some likely assumptions. IBP beef is featured at some leading Tokyo Western-style restaurants that seem to be on the cutting edge of the hospitality trade.

Western Meat Producers, Inc., at Pasco is now custom-feeding Angus cattle at their Hermiston, Oregon, feedlots, slaughtering in Pasco, and shipping split carcasses to their Japanese contractor. However, it is still unclear how Japanese grading standards are being applied to imported high-quality beef. Currently, planned Extension activity on the Pacific Rim will, among other things, address that issue. Western Meat Producers has also taken careful note of Japan's boxing preferences. There also remain remarkable factors to be considered by a select few cattlemen who might respond to this market in terms of the feeder cattle supply for Washington feedyards.

THE PERILS OF 1990

The steady increase in U.S. beef imports seen in 1989, the high cost of Wagyu feeders, the shortage of Holstein feeders, the excessive value of feed grains, the shortage of labor in the beef industry, growing consumer awareness, and the negotiated provisions of the U.S.-Japan Beef Agreement were all expected to produce an active trading climate in Japan during 1990. However, there have emerged several alarming factors that add risk for Washington producers involved in supplying the Japanese market.

On the positive side, early 1990 found the Japan Meat Trading Center Wholesale Cooperative Association making a test purchase of 30 tons of frozen beef from Tama Meat Packing, Inc., of Des Moines, Iowa. Nippon Meat Packers, Itoham, and Zenchiku are members of that cooperative. Sizzler Restaurants International signed Royal Company as its exclusive franchise operator in Japan. Five restaurants were planned as a pilot for eventual expansion. Upbeat reports on the impetus the Agreement has produced for free beef trade with Japan continued to find their way into trade publications during the spring of the year. Cargill reached agreement with Daiei to supply Japan's largest supermarket with direct shipments of beef cut to Japanese specifications. Under that arrangement, U.S. beef was to reach Japanese consumers 30 percent cheaper than domestic beef. Beef America reported a joint venture with the Jackson Company to sell specially seasoned frozen beef in vending machines throughout Japan at attractive prices for consumers. The "hype," along with associated innovations, was heating up.

However, academics and agri-journalists began showing skepticism by late spring. They reminded America that the Japanese are extremely exacting about beef quality, that the tariffs that follow liberalization will maintain imported beef prices nearly as high as they are with the current surcharges and taxes, and that there are probably outside limits to Japanese beef consumption under any circumstances. The gains made under the 1988 Agreement were pointed out as temporary. Some predictions began to appear that the U.S. will not benefit by liberalization over the long term and that growth of market share may have already peaked. News was then received that Australia had begun to introduce the Japanese-type meat-grading system for their grain-fed beef and that the Japanese themselves were collaborating in that effort to increase exports to their country. The LIPC also released frozen beef imported from Europe into the market in order to introduce and promote beef from areas with which the Japanese market is not familiar. Lastly, it was reported that the effects of the liberalized beef agreement would probably not affect Wagyu prices for up to five years after liberalization.

By mid-August 1990, faxed messages to WSU at Spokane indicated that the Kawasaki cold-storage facility of 42,000 metric tons was still operating at full capacity and that the Osaka facility, which began operations in July 1990, had already reached its 2,500 metric-ton capacity. August also produced a Knight-Ridder Financial News report–carried by *The National Livestock Weekly*–disclosing a sobering set of circumstances. In August LIPC, for the first time, bought less beef than it had tendered for. The lack of buying interest in Japan was attributed to glutted storage facilities and high retail prices. It is unlikely that the 1990 quota of 394,000 metric tons will be honored. At mid-year, Japan had purchased 2,500 metric tons less than in 1989.

Principal U.S. firms are realizing that the Japanese system is not getting the product sold and that the 1991 tariff of 70 percent holds out no relief, since the current tariff and user fee total a similar 70 percent. The cold-storage glut of U.S. beef is a reality, and not one expected to improve soon. Under the circumstances, American meats could begin to lose market share to Australia on price.

SUMMARY

Japan will continue to be an attractive market for responsibly produced, carefully marketed Washington beef. Respect must be paid to the sensitivities of the culture and the marketplace, and compliance must be meticulous. The U.S. share of the market may best be described as a niche within the B-grade spectrum for high-quality imported beef. Retail price and cold-storage capacity will continue to be limiting factors, and a consumption ceiling will eventually be reached. Both the IMPACT Center and Cooperative Extension can, and should, continue to play a useful role in these market dynamics.

Chapter 7

Japanese Demand for Vegetables and Vegetable Seed

Bill B. Dean

INTRODUCTION

The Japanese people utilize a large amount of vegetables in their diet. The traditional base food, rice, is supplemented with other foods such as fish, vegetables, and fruit. Increased use of bread, beef, and other products has served as an indicator of changes in dietary habits, the greater availability of products, and increased per capita income. These changes in eating habits include Western-style consumption of non-traditional foods as well as an increase in eating out–especially as a family unit and among young working women.

Japanese farmers are at a critical crossroads at this time. A high percentage of the farmers are over 60 years old, and the incentive for the next generation to take over the farm is not great. Increased labor costs and decreased availability of workers are also making it difficult for existing farms to remain competitive. The overall production of vegetables has remained relatively stable, but because of higher yields the area used for vegetable production has decreased consistently since 1970. There are many specific changes in production that have occurred as a result of competition from imports. Details about vegetable production can be obtained from a previously published report (Dean, Hasslen, and McCall, 1988).

VEGETABLE IMPORTS

A large selection of fresh and processed vegetables are imported by the Japanese each year. Imports originate in the United States,

59

China, Taiwan, New Zealand, India, U.S.S.R., Canada, Thailand, Philippines, Israel, Holland, Spain, Chile, South Africa, Belgium, Argentina, France, Italy, West Germany, Sweden, and Mexico. The United States occupies a significant position in imports to Japan, with an 18.7 percent market share. The majority of these products are of high per-unit value.

There are three major conditions that dictate success in this market. They are: (1) product quality, (2) availability, and (3) market acceptance. Within each of these areas, there are significant points that can be clarified.

Product quality includes several factors, and it varies depending on the market segment for which it is intended. In order to achieve success, the product must meet the high end of local quality standards and it must be consistent in meeting these standards. If there is no competition due to seasonal availability or sole-source marketing, standards may be different; however, consistency in meeting agreed-upon standards is still important.

The nature of fresh-vegetable production limits its availability by seasons which can be used as an advantage by prospective importers. Vegetables that are available during Japan's off-season are relatively easy to market. Knowing when the local production is available and when other potential import competition is available is essential to understanding the market potential. As Japan reduces subsidies, relaxes import duties, and eliminates quarantines, price competitiveness may significantly favor imports of some products even when local produce is available.

Market acceptance takes many shapes in Japan. Acceptance can be due to familiarity because of historical usage, to promotional effort and other marketing techniques, or to real or artificial controlling factors (such as quarantines). Plant quarantine restrictions are listed in Table 7.1. The market plan for each product needs to be carefully developed so that it identifies the appropriate market segment and/or allows for future expansion into other market segments. For example, if it is desired to eventually sell a product in upscale markets or as gifts, it may not be advisable to introduce the product in a common setting and at base values. It will be easier to reduce the price of an item after it is successfully marketed at the upscale level if common use of the product is desired.

Washington State farmers produce a wide variety of vegetables, as reported previously (Dean, Hasslen and McCall, 1988). There is potential for the vegetable producers to expand the number and variety of products they grow if the market demand is evident. Currently, in excess of 75 percent of the state's vegetables are marketed in a processed form. Proximity to markets, seasonality, and marketing priorities all influence the amount of vegetables entering the fresh market.

If we consider the trends in Japanese eating habits, our vegetable production and marketing system is ideally suited for this large market. The following scenario is intended as a futuristic glimpse of how Pacific Northwest producers, and Washington State vegetable growers in particular, fit into the agricultural export picture (primarily to Japan but also to the rest of the world). Vegetable crops are relatively easy to produce when grown in a suitable climate, are usually labor-intensive when produced for fresh market, and have a high value per unit of land area from which they are grown.

Vegetables grown for the fresh market will, in the long term, require production systems that provide (1) adequate land area and climatic conditions, (2) cheap, plentiful labor, (3) appropriate technology to provide a quality product, and (4) adequate handling and transporting systems to deliver the product to the market. This contrasts somewhat with vegetables grown for the processing market. The production system for processed vegetables must include (1) cheap, plentiful land with optimum climate for specific quality requirements, (2) cheap, plentiful labor, (3) appropriate technology to produce a quality product, (4) adequate storage or production capacity to create an inventory for year-round supply, (5) an adequate handling and transportation system, and (6) a cheap and readily available supply of water and electricity.

There is usually a difference in where the labor is utilized between these two production alternatives. Fresh packaging usually requires hand harvesting as well as packaging, whereas processed vegetables are usually harvested by machine. Processing plants usually require large amounts of fresh water and a cheap source of power for running equipment. The product quality may also be different between the two systems. Fresh produce must have a high cosmetic quality as its number one criteria, whereas the raw product

TABLE 7.1. Applicable Japanese Plant Quarantine Restrictions on Imports of Fresh Fruit and Vegetables from the United States (Excluding Hawaii)

Commodity	Status	Prohibited Reason by Code
Artichokes	Enterable	(A) Coding moth
Asparagus	Enterable	(B) Tobacco blue mold
Bell peppers	(B)	(C) Colorado beetle
Broccoli	Enterable	(D) Burrowing nematode
Brussels Sprouts	Enterable	(E) Potato canker
Cabbage	(C)	(F) Golden nematode
Cantaloupes	Enterable	(G) Sweet potato weevil
Carrots	Enterable	(H) Small sweet potato weevil
Cauliflower	Enterable	
Celery	Enterable	
Chicory	Enterable	
Corn	Enterable	
Cucumbers	Enterable	
Eggplant	(B)	
Endive/Escarole	Enterable	
Garlic	Enterable	
Green Beans	Enterable	

Green Onions	Enterable
Honeydews (Melon)	Enterable
Lettuce	Enterable
Mixed Melons	Enterable
Mixed Vegetables	Enterable in frozen form
Onions	Enterable
Parsley	Enterable
Parsnips	Enterable
Peas	Enterable
Potatoes	(E) (F)
Radishes	(D)
Romaine	Enterable
Rutabagas	Enterable
Spinach	Enterable
Squash	Enterable
Tomatoes	(D)
Turnips	Enterable
Watermelons	Enterable
Yams	(H)

for processing may have specific taste, texture, or other quality factors that are primary.

The vegetable industry in Washington developed around the processing industry's large expansion in the 1950s and it has been enhanced further by institutional food outlets and product users. At the same time, Japanese consumers have been modifying their eating habits and have moved in the same direction as United States consumers. Other developing nations, such as Singapore, have also followed this pattern and it is expected that many others will develop in that fashion during the next few decades.

PROJECTIONS AND OPPORTUNITIES

Japanese frozen-food production increased a dramatic thirtyfold between 1960 and 1970; it increased sixfold between 1970 and 1987 (*Tradescope*, April 1989). Production of vegetables for freezing was 30,000 tons in 1970 and it increased to 100,000 tons in 1984. Imports caused a reduction of 10 percent by 1987, and further reductions are anticipated. Imports now occupy approximately 80 percent of the market demand for frozen vegetables. Approximately 55 percent of the frozen imports come from the United States. Potatoes and corn account for 40.8 percent and 12.7 percent, respectively, of all vegetable imports. The United States provides 81.6 percent of the potatoes and 90 percent of the corn. The other major imports are: green soybeans from Taiwan (35,000 tons); garlic sprouts from China (4,000 tons); spinach (in small, widely fluctuating amounts); mixed vegetables; cauliflower; carrots; and green asparagus. One item for which there is market demand (but I am unaware of anyone processing it in Washington) is frozen pumpkin. This product is sold as cut chunks and is used primarily in the home.

As an import to Japan, frozen french fries from the Pacific Northwest are the biggest success story, their numbers having risen consistently during the past five years. Between 1987 and 1989, imports of french fries to Japan increased from 72,041 metric tons to 105,442 metric tons. Continued growth in this area is expected.

Sweet corn imports from the U.S. to Japan have also been successful. In 1987, 26,288 metric tons of frozen corn were exported to Japan, compared with 33,762 metric tons in 1989. Even greater increases occurred in canned-corn exports. In 1989, 41,035 metric tons of canned corn were exported to Japan, compared with 31,798 metric tons in 1987. Sweet-corn products are very important in the Japanese diet.

The fresh-vegetable market in Japan has not been a major target for Washington producers. While frozen and canned sales have steadily increased (and our market share has been significant), production seasonality, transportation availability, and risk factors have kept much of the vegetable shippers from entering this market. Table 7.2 presents an overview of the feasibility of shipping fresh vegetables to Japan. The following assumptions were made:

1. Products with a shelf life of two weeks or less must be transported by air.
2. Average seasonal prices do not fluctuate widely.
3. Quality of the product is acceptable.
4. The product must be able to absorb a 50 percent markup from the wholesale price to be viable.

Cabbage. Cabbage is a non-enterable commodity because of the Colorado potato beetle. However, if the status can be changed, this product would be viable during the months of July, August, and September.

Cauliflower. This product is acceptable if shipped by surface transportation during July, August, and September.

Cucumbers. Cucumbers are not of high enough value to absorb air-freight costs, which are probably necessary to maintain top quality.

Lettuce. Surface shipment of lettuce is economically profitable if the product quality can be maintained. Potential shipping months are: June, July, August, September, and October.

Onions (dry storage). Shipment of dry-storage onions is feasible by surface transportation, but is marginally profitable. When prices are high due to production problems in Japan, imports are increased. This crop has high-volume prospects but low per-unit profits.

TABLE 7.2. Average Value of Washington State Vegetables at the Wholesale Receiving Station in Tokyo Compared with the Average Retail Price for the Same Vegetable in Japan.

Product		April	May	June	July	Aug	Sept	Oct	Nov	Dec
Beans (string)	FOB Tokyo air				1.24	1.24	1.24	1.24		
	surface	NA	NA	NA	NA	NA	NA	NA	NA	NA
	retail	2.94	2.05	1.97	NA	NA	NA	NA	NA	NA
	net	–	–	–	–	–	–	–	–	–
Cabbage	air	1.15	1.15	1.15	1.15	1.15	1.15	1.15		
	surface	0.75	0.75							
	retail	1.10	1.17	1.49	1.96	2.20	2.20	1.46		
	net	-.05	+.02	+.34	+.71	+1.05	+1.05	+.31		
Cauliflower	air				1.00	1.00	1.00			
	surface				.29	.29	.29			
	retail	1.11	0.99	0.87	1.07	1.38	1.12	0.95		
	net			NA	.07-.81	.38-1.12	.12-.86	NA		
Cucumber	air	NA	NA	1.08	1.08	1.08	1.08	1.08		
	surface	NA	NA	NA	NA	NA	NA	NA		
	retail	NA	.92	.83	.82	.82	.96	1.13	1.31	
	net	NA	NA	-.25	-.26	-.26	-.12	+.05		
Lettuce	air			.94	.97	.97	.97	.97	NA	
	surface			.33	.37	.37	.37	.37		
	retail			1.06	1.05	1.12	1.11	1.41		

	net (surface)			.80	.77	.69	.69	.69		
Onions (dry)	air				.93	.93	.93	.93	.93	
	surface				.18	.18	.18	.18	.18	
	retail	.49	.44	.39	.40	.42	.43	.41	.42	.45
	net (surface)				.22	.24	.25	.23	.24	
Peppers (bell green)	air				1.12	1.12	1.12	1.12		
	surface				NA	NA	NA	NA		
	retail	–	–	1.27	1.36	1.32	1.28	–		
	net	–	–	–	0.24	0.20	0.16	–		
Pumpkin	air						.90	.90		
	surface						.15	.15		
	retail						0.67	0.65	0.79	
	net (surface)						.52	0.50		
Spinach	air		1.19	1.19	1.19	1.19	1.19	1.19		
	surface		NA	NA	NA	NA	NA	NA		
	retail	1.10	1.17	1.49	1.96	2.20	2.20	1.46		
	net	–	–.02	+.30	+.77	1.01	1.01	0.27		

NA = not applicable

Green Bell Peppers. This product is not enterable because of tobacco blue mold. It is also of marginal profit potential, since air shipment is probably necessary to maintain top quality.

Pumpkin. Pumpkins have good potential volume-wise and, if shipped by surface transportation, will return a respectable profit. Since there are many types of pumpkins, market demands should be evaluated thoroughly.

Spinach. Spinach is a relatively high value crop that would be profitable during July, August, and September. It probably would need to be shipped by air freight.

Melons. The average price for melons is listed by type below (from *Horticultural Products Review,* January 1990). Honeydew melon imports have risen significantly since 1986 (to 20,485 metric tons in 1988) and account for 72 percent of the melons imported to Japan.

Watermelon	131 Y/kg (1988)	$0.41/lb (145Y/$)
Prince Melon	340 Y/kg	$1.05/lb
Andes Melon	345 Y/kg	$1.07/lb
Ames Melon	329 Y/kg	$1.03/lb
Muskmelon	793 Y/kg	$2.49/lb

The off-season imports of honeydew melons from the U.S. have been reduced by 65 percent in 1989 as a result of increases in Mexican exports.

Vegetable Seeds. The U.S. exported $403 million in seeds in 1987-88 (U.S. Seed Exports, Foreign Agricultural Service [FAS], 1988), of which 38.4 percent were vegetable seeds. The dollar value of seed exports has been increasing steadily since 1980 (10 percent a year). Japan imported 13.8 percent of the U.S. vegetable-seed exports in 1987-88, with a value of $21.3 million. Sweet corn, radish, and spinach are the largest vegetable-seed imports from the U.S., with values of $7.2 million, $5 million, and $2.7 million, respectively. These are followed by beans and carrots, with values of approximately $0.5 million. Washington State vegetable-seed producers supply the majority of the radish, spinach, and carrot seed for this market as well as others. Growth in this area is not expected to be great if a continued decline in Japanese vegetable production occurs. Seed exports to other producing countries that will be supplying vegetables to Japan are expected to increase.

EXPORT EXAMPLES

Asparagus

Asparagus exports to Japan were studied beginning in 1986. Following the study, it was determined that Washington could expand its exports during a short market window in April and perhaps the first week of May. A market promotion program was initiated by using funds from the Targeted Export Assistance program and the Washington Asparagus Growers' Association. Bill Smiley and Walter Swenson, from the Washington State Department of Agriculture, coordinated the project, and they were advised by Dr. Bill Dean of the Washington State University IMPACT (International Marketing Program for Agricultural Commodities and Trade) Center.

Asparagus shipments were uncertain during the first part of the shipping period because of frost in the growing areas. Grade and size requirements were made available to the packers so that consistent quality could be packaged. Although some grade and quality problems did occur, sales increased to approximately 82 tons the first year. The program was continued a second year, with sales increasing slightly.

The market for Washington asparagus was presented with the following problems. First, the asparagus grown in Washington has a darker, more purplish color than that from California and Mexico, both of which preceded Washington in the market. Although it is not a major problem, the color did cause some hesitancy in sales. In years when California is late in the marketplace (as in 1990), sales from Washington suffer; if the main growing region in Japan comes on early, Washington is again put at a disadvantage.

Frozen asparagus has made significant gains with the advent of an IQF (individually quick-frozen) type of procedure, in which asparagus can be frozen and packaged as individual spears. Currently, demand is greater than supply for this product.

Sweet Onions

Japan currently supplies 92 percent to 94 percent of its domestic needs of dry onions (1.2 million tons). Although dry onions are produced in abundance in Japan, no "sweet onion" (such as the Walla Walla, Vidalia, or 1015Y) are produced. Since it is well recognized that the Japanese taste is for delicate or subtle flavors, a mild sweet onion may have a place in their diet.

In 1990, a test was conducted to determine if a market exists for this product. Thirty food-industry representatives took part in a tasting of Walla Walla sweet onions at the Agriculture Trade Office (ATO) in Tokyo. The purpose of the tasting was to determine if the onion was an acceptable product, could occupy a special place in the market, or was of no real value.

The results of the tasting were that the onion has at least two market potentials. The first potential is as a general-use onion, which must be price-competitive with the pungent onion sold at this time. Because the Walla Walla sweet onion does not have an extended shelf life, it must be shipped by air freight and is therefore not price-competitive. The second potential market is as a specialty onion for gourmet restaurants, upscale markets, or gift giving. The onion will need to be more consistent in shape and quality than how it is commercially sold in the U.S.; however, a significant market advantage (compared with domestic markets) may make it feasible to modify the current production/packaging system.

CONCLUSION

Japan will continue to increase imports and decrease production of vegetables. Washington is positioned very well to utilize production and processing technology to provide value-added products to the changing food industry in Japan. Marketing fresh produce will continue to be difficult because of Japan's domestic production and competition of others. We should concentrate our marketing efforts in products that require technical skill, large land areas, and low production costs. With the free market era we have just entered, competition from low-labor-cost producers will make it difficult to compete for markets of products that do not require a high degree of technical skill.

Chapter 8

Marketing Apples in Japan

R. Thomas Schotzko

In the not too distant future, the Washington State apple industry will be selling apples in Japan. With that prospect in mind, a trip was made to Japan during the fall of 1989 to observe production and marketing of apples. This paper discusses some of the major observations made during the trip.

First, it is important to recognize that Japan, like the U.S., is not a single market. Japan is comprised of many markets. Just as we have regional preferences for apples in the U.S., Japanese consumers are not all of one mind when it comes to apples. The observations by a foreign visitor regarding consumer preferences as portrayed in a Tokyo supermarket may be valid only for the service area of that supermarket.

The best recent example of marketing philosophies in Japan, as described in the U.S. press, is the case of 7-11 stores. The firm that owns the 7-11 franchise in Japan views each store as a market niche. There are thousands of these stores in Japan and each is connected by computer to corporate headquarters. The product mix of each store is tailored to the wants of the customers of that store. Each store is treated as a market niche.

Although not as sophisticated, the Japanese apple industry also attempts to market in a discriminating manner. Auction sale prices are monitored by the packing cooperatives every day, and shipment plans are adjusted on the basis of that information.

Apple production in Japan is no different than in Washington, since a range of qualities and sizes are produced. The Japanese thin more heavily and field-cull in response to fresh-market signals and a low-profit processing market. The fresh market wants well-colored, large apples. Yet, not all apples make number one grade. In fact, there appear to be four fresh-market grades. The fruit offered for sale in the upscale markets represents only a small fraction of the production. Other markets are available for the lower-grade fruit.

One packing house was visited during the trip. While it is not known how representative that firm is of the industry, it does provide an indication of the volume marketed by grade. The packout for the previous day at that house included 2 percent grade-one fruit, 21 percent grade two, 49 percent grade three, and 23 percent grade four. The remaining 5 percent were culled and sold direct to consumers off the front porch.

Since there are several different markets for apples in Japan, focusing on the price and quality of product being offered for sale in upscale markets in Tokyo can yield a very false impression. It does not portray the overall potential and range of market requirements for apples in that country.

Japanese business decisions are typically based upon much more information than is the case here in the U.S. This attitude also appears to prevail in apples for at least some firms. During a discussion with a regional produce buyer, the issue of soluble solids was mentioned. The buyer indicated that he received soluble solids information from the packing house whose fruit he was purchasing. His customers (consumers) preferred apples with 15 percent to 17 percent soluble solids. This may be an informational need unique to upscale markets only. However, both grade-one and grade-two apples were observed on the shelf of this market; since the Washington extra-fancy grade is probably equivalent to the Japanese grade two, this means Washington apples may be able to compete in that market.

Depending upon the marketing strategy of the individual Washington packing house, changes in container size may be necessary. The packed cartons observed at the Japanese packinghouse and at the auction in Tokyo contained either two or three trays of fruit. Each tray contained 5 kilograms of fruit. Washington cartons currently contain 42 pounds (19.1 kilograms) of apples.

In Washington, packed-carton weight is the common unit of measure. However, tray weight for the larger fruit in Washington cartons is only about a half-pound less than the 11 pounds used in Japan. Firms with electronic weight sizers would be able to make minor adjustments and meet Japanese requirements with half-cartons.

Research on the Japanese apple market suggests that access to that market will not have a major impact on prices in Washington. However, Japanese consumer preferences for apples lean very heavily toward large fruit. The packinghouse visited during my tour did not pack any apples smaller than size 88 (approximately half a pound in weight). This penchant for large fruit will likely have a significant positive effect on prices of large fruit in Washington when that market is opened up. If Japanese consumers are willing to accept only Washington extra-fancy apples (and nothing smaller than a size 88), shipments to Japan could remove as much as 15 percent of the product meeting those specifications in a typical year.

Two other observations deserve mention. First, the Japanese do not wax their apples. They have found that shelf life of waxed apples was less than that of unwaxed apples. It would be in the Washington industry's best interests to work closely with Japanese buyers on this issue, to overcome the bias against waxed fruit. Further evaluation of consumer attitudes regarding wax is warranted. Obviously, it doesn't make sense to educate retailers if consumers really don't want waxed apples.

The other factor whose importance is unknown is shape. All apples observed during the trip were round, including the Starking Red Delicious. While the knobs at the calyx end of the Starkings were plainly visible, these apples were uncharacteristically round (that is, compared to Washington State Starkings). Some thought should be given to Japanese attitudes regarding shape, in order to (1) determine attitudes and (2) take advantage of the difference in attitudes regarding shape, if significant.

When the Japanese market opens up, there will be some sales of Washington fruit. In the short run, the amount of success will be determined by the willingness of our industry to learn about, and work within, that marketing environment. The accumulation of three sets of information now would improve short-run prospects for sales to Japan. First, there are auction markets throughout Japan, and daily price information is available on sales of domestic fruit. It seems safe to assume that grade and size information are also available. That information would help in identifying potential markets.

There is one potential problem with the auction data. At least in the Ota market in Tokyo, the highest-quality products do not reach the auction floor. Those products are diverted somewhere between the receiving dock and the auction floor. Prices paid for that produce is not reported. Fortunately, there appears to be sufficient volume to have confidence that the reported prices reflect actual market conditions and not just the price for rejected loads.

A tour of selected markets outside Tokyo to observe fruit quality and size would help individual packinghouses target the most appropriate markets for their products. That tour should include auction markets, retailers and neighborhood greengrocers, and contacts with potential buyers.

Grade standards are employed in Japan, but it was difficult to obtain much information on grades. A survey of grades would be valuable in evaluating the price data collected in Japan. Further, it would allow Washington marketers to describe offerings in terms of Japanese grades. This last point is subtle, but may have the potential to generate the greatest profit. Japanese importers take considerable advantage of their knowledge of local markets. Two examples immediately come to mind. The first involves Azuki beans, as reported by Dr. Tom Lumpkin at the IMPACT (International Marketing Program for Agricultural Commodities and Trade) Center's 1989 Symposium. He described how the Japanese importer wanted to pay $300 per ton less for Washington-grown Azuki beans than was being paid for Taiwanese beans and $600 less than was being paid for Japanese-grown beans.

The other example comes from asparagus. When the Japanese initially began buying fresh asparagus from Washington State, they were willing to pay $4 to $5 more than the current local F.O.B. market price. Japanese importers soon realized they could make their purchases at the local F.O.B. level, thus keeping for themselves the additional money they had been willing to pay earlier.

Within the next few years, Washington apples will be allowed into Japan. The impact of that access will be determined by the willingness of the Washington State apple industry to learn about, and take advantage of, the market niches that exist in that country.

Chapter 9

An Overview of the Washington and Japanese Wine Markets

Suzanne Callender
Raymond J. Folwell

Washington State has enjoyed success with its relatively new wine industry. While other U.S. wine regions are experiencing declining sales, Washington wine sales are increasing. From 1987 through 1990, combined in-state and out-of-state sales of Washington wine increased 18 percent. During this same time period, the sales of all wines in the U.S. have decreased.

Washington wine-grape acreage has grown from less than 2,000 acres in 1974 to 10,169 acres in 1988. This growing industry has been a benefit to the Washington economy and to the tourist trade as well. Annual retail product value is over 100 million dollars, and more than 500,000 tourists visit the wineries each year.

From 1986 to the present, per capita adult and total wine consumption levels for all of the U.S. have decreased. The top consumption states have seen a drop in annual per capita consumption levels. It is interesting to note that Washington has been in the top five per capita consumption states since 1983. As Washington wine production increased, producers were dependent on the Washington and Northwest regional market. While the in-state market share held by Washington producers has increased to 16 percent in 1990, it is increasing in a shrinking market. In 1990, slightly more than 50 percent of the total production of Washington wines were sold out-of-state. If the industry continues to grow at the rate it has in the past, the state and regional markets will become saturated and it will be necessary to look for additional market outlets, including international markets.

The most basic action that Washington wine producers can take to avoid an oversupply and lower price for their product is to concentrate on new markets. While this is a costly endeavor to undertake, the benefits in the long run will ensure economic stability and marketability for their product.

The Pacific Rim and Japanese wine market was the focal point of this study. Although the Japanese consumption of wine is very low in comparison to other developed nations, its rate of change is positive and large–about 13 percent average growth rate in the last three years in total wine consumption. The current per capita consumption level is about 0.9 liters. Some analysts for the wine market believe that the Japanese market for wine could triple in ten years. With this in mind, and given the fact that the Japanese production capacity for wine can only meet about 50 percent of Japanese market demand, most market analysts consider Japan a good market prospect.

The Japanese market has unique characteristics that economically, culturally, socially, and politically affect it differently from any other market system. Economically, the average Japanese consumer has more disposable income than that of any other consumer in the world. This makes the Japanese market lucrative for high-quality and luxury goods. Culturally, the Japanese do not have any historical wine practices, except with sake (Japanese rice wine). This could be either a hindrance or an advantage, given the market's receptiveness to new products. Socially, the Japanese market tends to be very homogeneous in its consumption and habit patterns. Trends and fads are very important in Japan and do not have as short a lifespan as in other countries. Politically, Japan's minority of farmers have more power than city dwellers, who are the majority of the consuming market. This creates more protectionist agricultural practices than would likely occur if political power were equally distributed.

There are two very different product groups in the Japanese wine market: domestic wine and imported wine. The majority of domestic Japanese table wine is not varietal but generic. In 1989, about 120,000 metric tons of grapes were used to make all of the domestic wine in Japan. Of that 120,000 metric tons, only 30,000 metric tons were grown in Japan. The rest was imported as fresh grapes, grape must, or bulk wine. In order for domestic wine to be called "Japanese," only 25 percent of the wine needs to be from Japan. To be called a varietal wine, a minimum of 75 percent of the variety must be in the wine, regardless of where it was produced.

Most of the Japanese grape acreage is not planted in typical wine varieties, such as those found in premium wine, but rather in varieties that can be used as fresh table grapes. Since the majority of domestic wines are blends, they are given a packaging name. The large alcohol companies are especially adept at marketing these blends. Very innovative packaging and promotional techniques are being utilized in order to create niche and tailored markets. Wines are labeled with specific descriptions of their characteristics–such as "dry," "fresh," or "light wine for drinking with traditional foods"–to give the consumer knowledge for decision making. Instructions are also included on the back as to serving and proper wine etiquette. All of these innovations are intended to help sell a product that is not very well known within the populace.

The target markets for these wines are varied. Medium and small wineries concentrate on tourists and point-of-sale, since the majority of their product is sold at the winery. The larger alcohol conglomerates, which also have wineries, do have a tourist trade, but the majority of their business comes through distribution in the metropolitan areas.

Since the majority of Japanese are not familiar with wine, the domestic wineries have taken a promotional stance to familiarize the population with wine and thus create a less threatening image. The majority of their product lines are focusing on the less expensive, lighter-tasting introductory wines that would be attractive to an inexperienced palate. For the intermediate wine drinker, there are product lines that are medium-priced and have a higher-quality image. These products come in the form of varietal and generic table wines, with higher-quality packaging and product to suit the more knowledgeable consumer of wine. There are also high-priced, high-quality specialty products that some wineries produce, such as "late harvest" or other differentiated products.

The smaller, family-operated wineries usually have one or two products. The larger wineries have larger product lines. Usually, as in the case of Sapporo and Suntory, these wineries have special relationships with out-of-country wineries that provide or help produce some of the higher-quality products. This type of symbiotic relationship is advantageous for the company promoting a complete product mix and it also gives the out-of-country product instant brand affiliation and access to marketing channels. This is a very popular marketing strategy used by foreign companies to access the Japanese market. This also helps the Japanese company, since there is a perception in Japan that imported wines are higher-quality wines. With this perception, Japanese companies improve the overall image of their product mix and thus improve their domestic brand's image. However, these wines are still considered imported and would be considered competition for the wine consumer's dollar.

Imported wine is brought in by both trading companies and importers. As stated above, foreign wines are perceived to be of higher quality or as having "snob" appeal within the Japanese market. Of all of the foreign wine being brought into Japan, France has the largest market share (47.3 percent in 1989), followed by Germany with 22.4 percent and the U.S. with 15.4 percent (Figure 9.1). Italy, Australia, Portugal, Spain, and others trailed with 4 percent or less each. French wine serves as the epitome of a marketable wine product in Japan.

In the 1988 to 1989 period, the volume of French wine imported by the top-thirty trading houses increased from 1,509,600 to 2,014,750 liters–a 33.5 percent increase. Overall volume increases for French wine imports in the same period was only 1.9 percent. This indicates a shift in market share from small importers to the larger trading houses.

Smaller trading houses that import French wine cannot compete in the market channels, points-of-sale, and promotional arenas of the large alcohol companies. The smaller trading houses have focused on premium French wine markets, which have a smaller consuming population but a higher profit margin. Currently in Japan, this market is expanding due to increases in disposable income and to consumers' increased exposure to these wines while traveling in France. (France is the top destination for Japanese travelers in Europe.)

Image is an attribute that cannot be clinically analyzed, yet it can

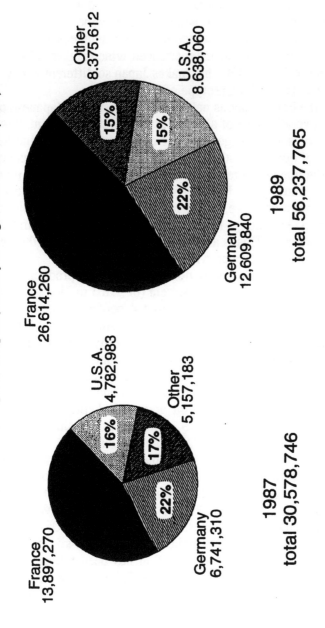

FIGURE 9.1. Wine Imported in Japan by Country of Origin, 1987 and 1989 (Liters)

Other
8.375.612

U.S.A.
8.638,060

15%

15%

22%

France
26,614,260

Germany
12,609,840

1989
total 56,237,765

U.S.A.
4,782,983

Other
5,157,183

16%

17%

22%

France
13,897,270

Germany
6,741,310

1987
total 30,578,746

Source: Japanese Ministry of Finance, 1989

be a critical point in the marketability of a product. This is where the French excel with their wines. French wine, all around the world, evokes a certain reverence that is instantaneously acknowledged by experienced and inexperienced wine drinkers alike. It is no different in Japan. But what makes Japan so different is how the populace reacts to this stereotype of French wine.

Due to the homogeneous nature of the consuming Japanese populace, the overall image of a product type is very important. It is the first stage of a knowledge base for a product. French wine has a reputation of quality, product assurance, and value. This reputation conveys the image of prestige and knowledge of wine, and it commands a premium price. French wine has an image that is a positive, or at least not a negative, factor in the consumer's wine consumption decision. This image has definite advantages, as seen in potential profit margins. This is another fact that could explain the rise in volume of imported French wine.

The U.S. wine industry is dominated by one producing state, California, which produced 93 percent (or 396.3 million gallons) of all U.S. wines in 1989. Of this, about 4 percent was exported to other countries. Since California wines constitute the bulk of total U.S. wines, California is the only state explicitly identified in export statistics. Very little is actually known about other states' wine export volumes, since no data has been published on such movements. Due to the lack of data, no specific conclusions can be made for wines from small production areas. Marketing statistics collected from Japan will be on U.S. production only, since there is no distinction in Japan between U.S. wine regions.

In 1989, Japan was the most valuable overseas table-wine market for the U.S., with $17.2 million in sales. On a volume basis, Japan was a distant second (with 2.9 million gallons) to Canada, which was a 4.3-million-gallon market for U.S. wines. The average price of U.S. wine sold to the Japanese was $5.93 a gallon. Of all the U.S. export markets, Japan has bought the most expensive U.S. wine. However, the average value per gallon has been falling since the high value of $6.22 per gallon set in 1987. Several factors could be affecting this trend: (1) the price of U.S. wines is correlated to the fluctuation of world currencies, (2) U.S. wine supply and demand could be forcing the prices downward, and (3) the U.S. could be sending Japan less expensive wine to increase interest from Japanese buyers.

As a nation, Japan is one of the few developed countries that has an increasing consumption of alcohol: 1 percent average annual growth during the 1985-90 period. It also is one of the few countries with increasing wine consumption rates: 13 percent annual growth during the same period. Due to its domestic production situation (as discussed earlier), Japan will always be heavily dependent on imports of wine. This is fueled by the image that higher-quality wine comes from out-of-country sources. The 1990 projection is that 55.5 percent of all still wine consumed in Japan will be imported bottled wine, not including bulk wine. Imported bottled wine experienced a 26 percent total average annual growth in the 1985-90 period.

Despite these excellent market conditions, the U.S. and Washington State wine producers must deal with a few hurdles in the Japanese market. First, and foremost, is the decreasing market share for U.S. wines in the 1987-89 period. Average annual growth for these years for total wine import sales was 35.6 percent. U.S. sales grew by 34.4 percent, a 0.2 percent loss of market share for the U.S. The other two largest competitors, France and Germany, increased their market share by 1.9 percent and 0.4 percent, respectively. What is most disturbing about this situation is not the relatively small decrease in market share for the U.S. but the increase in market share of its major competitors. These two major competitors had a substantial 69.7 percent share of the 1989 Japanese import market. The 30 largest handlers of U.S. wine in Japan have also decreased their volume by 8 percent in the 1989-90 time period. If these trends continue, the U.S. will lose its market share.

Another major hurdle that the U.S. wine industry has to be aware of is its image in Japan. As compared to wines from France and Germany, U.S. wines tend to be less highly regarded. Wines from France and Germany have a high price-to-quality image within the Japanese consumer's mind. U.S. wines have not been able to match this image. This is due to a mixture of components that have to do with price, packaging, promotion techniques, and image building.

French and German wines are highly visible because the Rhine, Mosel, Bourgogne, and Bordeaux production regions have long been famous for their wine production. This familiarity is reinforced by packaging and presentation of the French and German wines that is much different than for U.S. wines. The U.S., being the newest wine production area of the three, does not enjoy this established image as a wine-producing region. U.S. wines do not emphasize their region of origin in their packaging as do their European competitors. The quality image is enforced also by the regulations the French and Germans impose on their wine-growing regions, to ensure the consumer of the quality of the product they are purchasing. The consumer is less assured of product specifications when purchasing a U.S. wine.

Probably the most critical impediment for U.S. wine is its past product image. In the 1970s and 1980s, the U.S. wine industry was producing for the less expensive table-wine market. These markets have slowly dwindled and now the emphasis is on premium varietal and generic table wines. The French and the German wine regions have always had product lines to fulfill the premium market demand. Their penetration of the Japanese market was easier due to these premium table-wine product lines. The U.S. was not prepared for this type of trend, since the majority of its production was not intended for the premium table-wine market. During the late 1980s, table wines from the U.S. have enjoyed great success in the marketplace. However, the past image remains and it can be a hindrance when trying to compete against countries that do not have the same history.

At this time, U.S. wines are in an uncertain position within the Japanese wine market. Ranked third in overall import volume, suffering from a possible image problem, and being outperformed in market-share growth by their strongest competitors, various segments of the U.S. wine industry need to evaluate their situation and develop remedial market strategies. The newer U.S. areas of wine production should be aware of the situation within Japan and determine if their products can be in a position to be sold within this market.

Washington's wine industry has some characteristics that set it apart from all other wine areas in the U.S. In 1987, Washington produced 3.9 million gallons of wine, about 0.8 percent of total U.S. production. Washington enjoys a limited but good reputation as a wine-producing region within the U.S. but has no established image in Japan. Even with such a small output relative to the U.S. total, Washington produces the second-largest amount of "premium table wine," after California.

"Premium" wine has a very vague definition. There are no specific quality criteria set to distinguish premium wine from non-premium wine. It is more a matter of perception. Premium wines in the U.S. are generally varietals that have a higher image due to their quality, packaging, pricing, and presentation of product to the consumer. This type of product also commands a higher price. The premium table-wine market, along with that of champagne, is the only wine segment that is increasing in the U.S. and in other countries. About 80 percent of Washington wines fall into this premium wine category. This explains why Washington's wine industry is experiencing growth in sales of its products while total U.S. wine consumption is decreasing. This type of wine is also an increasing segment of the Japanese market.

Washington has one large winery that accounts for about 40 percent of Washington's total production. The rest of its production comes from numerous small- and medium-size wineries. Washington is a young wine region, and the majority of U.S. wines have historically come from California. This anonymity could be an advantage or a disadvantage. Washington could build a reputation distinct from the U.S. image and present itself as the premium wine region in the U.S. This could help bypass the past image that has plagued U.S. wine within the Japanese wine market. The limited wine production within the Washington wine region would help to facilitate this premium image, if promoted correctly.

Another characteristic that Washington State has is the distinctiveness of its wines. Varietal flavor and acid content differentiate Washington wines from other U.S. production. No data was found in Japan that would give any indication of taste, price, and quality combinations that various market segments would prefer; however, with the information that has already been discussed, certain segments of the Japanese wine market can be looked at as possibilities for targeting with Washington wines (Figure 9.2). The largest market share was in the $10.20-$13.65 retail price bracket in the Japanese market, which captured almost 50 percent of total sales. The

FIGURE 9.2. Japanese Wine Market by Price and Market Share and by Market Outlet, 1988

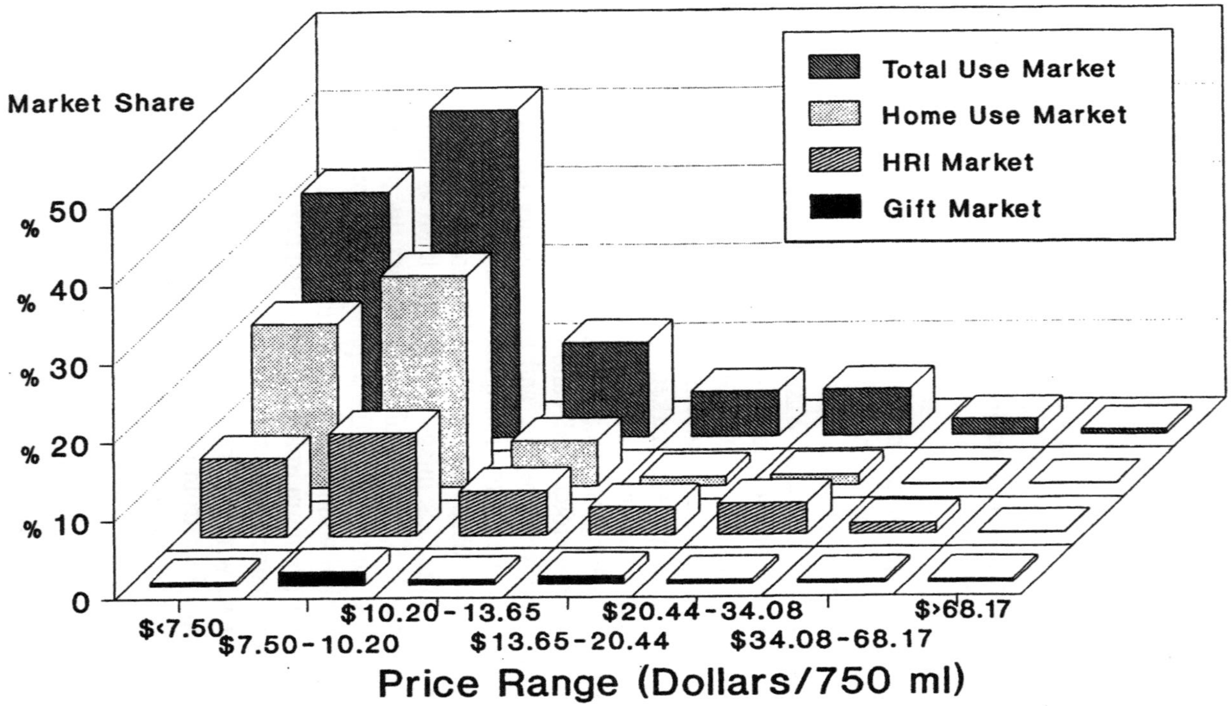

Source: WANDS Wine Trade Magazine

86

largest portion of the market in the lower price segments was for the home-use market. The higher-priced segments sold better in the hotel, restaurant, and institution (HRI) and gift markets. As for imported wines, 42 percent were sold in the $7.48-$10 price range. Higher-priced wines had a good proportion of market share as well.

The gift-market segment is an interesting market segment, but it only accounts for 5.6 percent of the total market in Japan. However, when compared to the other segments within price ranges, its market share increases with price. That is, as the price of wine increases, so does the importance of the gift market. An example is the 10.2 percent market share that the gift segment has within the $20.44-$34.08 retail price bracket, whereas HRI has 66 percent and home use, 22.8 percent. In the $34.08-$68.17 retail price bracket, the gift-market share jumps to 23.8 percent, with HRI at 68.2 percent and home use at 7.7 percent. This type of information can be very beneficial when identifying market niches for Washington State wines to penetrate based on price within the Japanese market.

Washington State has some advantages over the other U.S. wine-producing regions. It has, for instance, a differentiated premium product that can be packaged with a high quality-to-price ratio. There is no negative image that would require effort to reverse.

Production is limited, and this contributes to the product's exclusiveness. However, with the last few advantages there are a few difficulties that arise. The first, and foremost, is building an image of Washington wines within the Japanese market. Creating an image is a large and expensive task, particularly within Japan. Like the U.S., Japan is a highly targeted consumer society. Because of its level of disposable income, Japan is saturated with advertisements. To create a Washington wine image would demand a commitment of capital and time, something that the Washington State wine industry might not find appealing at this time. There is little hope that any promotional or advertising effort by the average individual winery would have enough impact on the market on a countrywide basis. This is why promotion as a region would make the most sense. Capital and effort could be pooled, resulting in building a broader market base from which all producers could gain. Pooling of resources would also offset Washington's other main problem—the small volume targeted for the Japanese market.

The other, most pressing, problem would be maintaining a consistent supply for the Japanese market. To build a stable market, the supply into that market must be steady. The Japanese consumer is far too sophisticated to be committed to an inconsistent market supply. There are too many other products that could be used as substitutes. The biggest complaint that was voiced by members of the distribution channel in Japan was inconsistent supply from wineries.

Many brokers are cautious about dealing with smaller wineries because these have been unreliable supply sources in the past. This need not be the case for Washington wineries. The larger wineries will find it easier since they can probably produce enough volume to make the venture profitable for a broker. What makes a winery "large enough" depends upon the broker. The majority of the wineries in Washington State are probably not big enough to deal with brokers on an individual basis. Again, pooling of resources might be the answer. Smaller wineries could pool their volume and pool their promotional dollars in order to benefit their general region and their individual winery as well.

Unfortunately, this pooling of resources might not sound appealing because of the lack of clear differentiation between wineries. The fact remains that there are many wines in the Japanese market that are bought simply because they come from a well-known region. There are fewer wines in Japan that are bought because they come from a specific winery. Only the few world-renowned wineries can command such brand loyalty for their product (e.g., Chateau Rothschild, Margaux, and Beaumont). Right now, however, Washington State wines do have characteristics that could be easily marketed as coming from a distinct growing region.

Washington State stands in a good position to penetrate the Japanese market. The products coming out of Washington satisfy the more educated wine consumer in Japan. This type of consumer is more adventurous in trying new products and would appreciate the quality of Washington State wines. These consumers are willing to pay higher prices for their purchases. Accurate demographic data on how many consumers fall into this category were not available, but this market segment is thought to be increasing. Enrollment in wine clubs and participation in wine salons are increasing, as is the acceptability of drinking wine with meals.

To penetrate the Japanese market, Washington State wineries must coordinate their efforts in order to increase their market power. Most Washington State wineries do not have the volume to blanket a market to create consumer loyalty. But together, as a region of wine production, Washington wineries can pool their knowledge, wine production volume, and capital to build a regional image. This image could function just as the regional images of the French and German wines do in promoting sales.

Chapter 10

What If Japanese Wheat Imports Were Liberalized?

Danna Moore

INTRODUCTION

Japan is Washington State's largest wheat market and is also the most important international market for white wheat's end-use characteristics for manufacturing. Japan's trade in wheat is controlled by the Japan Food Agency (JFA). The JFA imposes trade barriers that have been criticized as being as restrictive as quotas (Australian Bureau of Agriculture and Resource Economics, 1988). While the JFA has not been addressed under the General Agreement on Tariffs and Trade (GATT), it is a trade distortion that violates the code of the Uruguay round. Elimination of the JFA and/or relaxation of trade restrictions in Japan for world trade in wheat would reorganize the wheat industry and wheat production in Japan as well as shift buying patterns of wheat classes purchased. It is hypothesized that wheat class purchases would align toward the best end-use properties.

Understanding Japanese demand for Pacific Northwest-produced white wheat involves an understanding of many aspects of the world wheat trade. World wheat trade has intensified with increased recognition of differentiation in important wheat quality characteristics. Overall Japanese wheat-product consumption has stagnated, with a shifting emphasis on quality directed to specific end uses. Japanese wheat buying is government-controlled (by the JFA) and is dependent upon the amount of highly subsidized Soft Red Winter

Wheat produced in Japan. The Pacific Northwest and, specifically Washington State, produces Soft White Wheat that is exported to Japan. Hard White Winter Wheat is a newer variety that has not been significantly produced in Washington State or in the United States. Any production of Hard White Winter Wheat has important consequences for Pacific Northwest-produced white wheat, since blending is a common practice through the marketing system and wheat shipments to Japan pass from Pacific ports.

WORLD WHEAT TRADE WITH JAPAN

There are many different classes of wheat produced and traded on the world market. Differences among classes are attributable to indigenous and extraneous factors (Wilson, 1989). Indigenous characteristics include color, protein levels, quality, strength, and hardness; they are the result of environmental conditions, breeding programs, and varietal selection practices. Extraneous characteristics are the result of a particular country's marketing system and they include grading, sprouted wheat, non-millable materials, blending, and shiploading regulations. Both components determine quality and are important competitive factors in trade with Japan. As intensity of competition between suppliers increases, so does the importance of differentiation between wheat-class quality.

Price is an important indicator of quality-related differences between suppliers of wheat to Japan. Wheat, even if of the same type, grown in a different country is not identical and will possess characteristics that influence its end-use performance for specific manufactured products (U.S. Congress, Office of Technology Assessment, 1987, 1989). Most countries that supply wheat export predominantly one wheat class. The U.S. is the only country that exports significant amounts of more than one class. Hard Red Winter is the most abundantly produced and exported wheat class from the U.S., followed by Hard Red Spring, Western White, and Soft Red Winter.

The quantity of protein in wheat serves as a proxy for protein quality and affects the value of wheat traded across sources and through time. Protein quantity and quality affect the gluten strength of doughs. High-protein wheats are highly desired for bread production and soft low-protein wheats are used for crackers, biscuits, and confectionery products.

There are several important impacts as a consequence of differing country policies. Uniformity of wheat produced from crop year to crop year–across production regions and by shipload–has a direct influence on end-use performance in wheat manufacturing. Importers of wheat into Japan have criticized wheat from the Pacific Northwest (and other regions of the U.S.) because it lacks uniformity. For example, they specifically claim that both Western White and Soft White wheats have protein contents that have varied widely and have trended upward. (Protein content of Western White to Japan should be 9.5 percent or lower.) Consistency between cargo loads for all quality factors (protein, falling number, dockage, etc.) would improve baking properties of Western White (Washington Wheat Commission, 1988).

Prior to 1973-74, all wheat-class prices were increasing with little variation of price by wheat class, reflecting the relative world shortage of wheat supplies. Wheat prices peaked for all wheat classes in 1973-74 and again in 1981-82, and have since declined. Of particular interest are the delivered prices of wheat classes C and F in Japan, relative to U.S. Ordinary Hard Red Wheat, which is the U.S.' most abundantly produced and exported wheat class. The ratio of specific wheat prices to that of Ordinary Hard Red Wheat is shown in Figure 10.1.

Price differentials among wheat classes and origins were small prior to 1974, thus ratios of specific wheats to Ordinary Hard Red Wheat were close to 1:1. Since that time, price differentials have increased between wheat classes traded, especially for higher, stronger protein wheats. Canada Western Red Spring (13 percent protein) from Canada and Dark Northern Spring (14 percent protein) from the U.S. have traded significantly higher than Ordinary Hard Red Wheat. Appreciation for Canada Western Red Spring (13 percent) relative to Ordinary Hard Red Wheat has been greater than that for Dark Northern Spring 14%, most likely reflecting the superior bread-baking qualities of Canada Western Red Spring (13 percent). Australian Prime Hard (13 percent) from Australia, while trading at a higher price than Ordinary Hard Red Wheat, did not

FIGURE 10.1. Ratio of Wheat-Class Import Prices to HRW Ordinary, C and F, 1972-1987

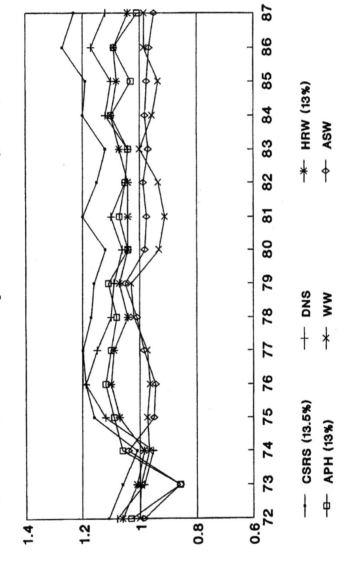

— CSRS (13.5%)   + DNS   ＊ HRW (13%)

↔ APH (13%)   ＊ WW   ◇ ASW

Ratios of Various Wheat Class Prices to HRW (ord). C&F Japan US$ per ton.
SOURCE: International Wheat Council

94

appreciate relative to Canada Western Red Spring (13 percent) or Dark Northern Spring (14 percent). Australian Prime Hard (13 percent) traded very close to the price of Hard Red Winter (13 percent).

Both Soft White Wheats, Western White and Australian Standard White, traded below Ordinary Hard Red Wheat, except for crop-years 1978 and 1979. The largest differential between Western White and Australian Standard White occurred in the 1980-82 period, with Australian Standard White capturing some premium over Western White.

The existence of price differentials for each wheat class suggests that increased ability to further differentiate will increase competitiveness of a wheat class marketed. While country of origin is important, other country-specific characteristics are potentially more important. As baking technology becomes more sophisticated, importing countries demand uniformity in the factors that affect baking performance.

FACTORS ACCOUNTING FOR WHEAT-CLASS DIFFERENTIALS TO JAPAN

Japanese Consumer Wheat Product Demand

The sociodemographic characteristics in Japan are changing dietary requirements and the overall demand for food products. For the last 20 years, Japanese families have experienced an increase in economic well-being, with increased personal disposable income and decreased food expenditure as a portion of the household budget (Annual Report on Family Income and Expenditure Survey, 1961-87). Other changes include more women entering the work force (which also contributes to increased household income), the increased cost of having children, and the demand for products and services that offer convenience. The average age of the head of household has increased, which translates into both dietary preference changes and increased ability to purchase. With these changes in household makeup have come changes in consumption of wheat products.

Wheat-flour consumption within a household usually declines

with increasing personal incomes. For Japan, wheat-flour consumption at home has been consistently low (less than 0.1 percent of food expenditures). Not much food prepared at home uses wheat flour.

Bread expenditure per household increased steadily between 1967 and 1987 (Figure 10.2). White bread expenditure increased until 1981 and has since declined, whereas expenditure on other breads continued to grow. Statistics on bread quantities purchased and on prices show that increasing price contributed considerably to increased family expenditure on bread. White bread quantities purchased increased until 1981 and have since declined. Other breads, such as sweetened, have experienced increased quantities purchased.

Expenditure on all noodles showed a slow steady increase from 1963 to 1987 (Figure 10.3). From 1973 to the present, total noodle expenditure tripled. Further division of noodles into boiled, instant, dried, and other allows for distinguishing the fastest-growing noodle market for Japanese households. Instant noodles accounted for the largest portion of noodle expenditures. This is also the most convenient and expensive form of noodle product. In 1987, all noodle categories approached equal expenditure shares and equal quantities in the family budget. The annual average quantity of all noodles purchased by Japanese households has remained between 3,600 and 3,900 kilograms. Further categorization shows a decline in boiled noodles, a slight increase in dried noodles, a steady quantity of instant noodles consumed, and a decline in the quantity of other Chinese noodles consumed. The most expensive noodle is the instant noodle, and this is the one with the lowest quantity consumed. Boiled noodles showed the lowest price and the highest quantity of noodle purchased. There is a small difference in quantity consumed between the noodle categories; the approximate difference in 1987 was 400 kilograms. However, instant noodle price approximated three times that of boiled noodle price in 1987.

. Average household expenditure on cake and confections has increased from less than 20,000 yen in 1965 to over 70,000 yen in 1987 (Figure 10.4). Each cake category showed expenditure increases. Japanese Fresh Cakes are the most rapidly increasing of the cake types reported. Kasutera-style Japanese cakes were the lowest expenditure category and have remained relatively stable. The lack of information on price and quantity prohibit speculation as to whether consumption has changed or whether increased prices are contributing to increased household expenditure.

Consumption of wheat-based products in Japan has influenced the mix of wheat classes purchased. Each of these wheat product categories is dependent upon different wheat classes to obtain the attributes of protein, gluten strength, and color necessary for each manufacturing process and final product quality. For bread baking, as preferences shift from the U.S.-style Pullman loaf to other specialized breads, the demand for wheat like Hard Red Winter will change. Current demand for wheat classes has shifted to those classes that deliver gluten strength and higher protein content to maintain loaf volume for specialty breads. The role of Canada Western Red Spring has become more prominent in trade, since its protein strength and quality are superior in end-use performance (Dick et al., 1986).

Confectionery production in Japan is reliant upon Soft White Wheat. Flour millers consider the low-protein soft gluten content of soft wheat to be essential to the baking quality of sponge cakes, cookies, and other confectionery products. Acceptability of final confectionery products is also highly dependent upon the amount of sprouted wheat, which directly affects alpha amylase activity of wheat flour. These are factors highly influenced by a country's breeding, selection, and blending practices (U.S. Congress, Office of Technology Assessment, 1987).

Noodle consumption in Japan has shifted from consumption primarily of Japanese wet noodles to consumption of almost equal amounts of Japanese wet noodles, Chinese noodles, and instant noodles. The preferred wheat class for Chinese- and instant-noodle production is Australian Standard White. While other wheat classes can be substituted in the noodle flour mix, only Australian Standard White gives consistent noodle quality (as measured by bite and color). Hard Red Winter classes from the U.S. are the least preferred wheat in noodle manufacturing. At present, the U.S. does not produce a White Wheat class that competes with Australian Standard White in Chinese- and instant-noodle manufacturing. As noodle consumption has shifted, wheat-class preferences have

FIGURE 10.2. Annual Average Household Bread Expenditure, Price, and Quantity in Japan, 1965-1987

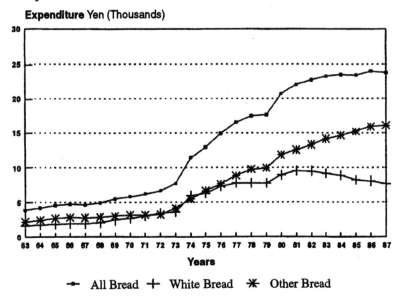

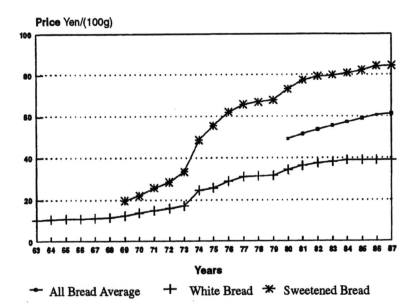

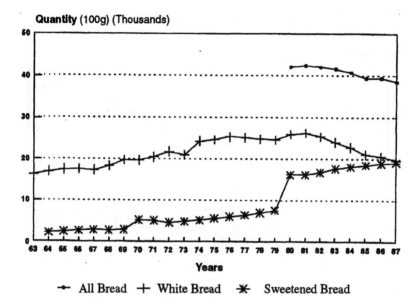

Quantity (100g) (Thousands)

Years

◆— All Bread +— White Bread ✱ Sweetened Bread

. Japanese Statistics Bureau. Family Income Survey.

shifted from utilizing mostly Japanese Soft Red Winter and Hard Red Winter to utilizing higher levels of Australian Standard White.

WHEAT MANUFACTURING CHARACTERISTICS AND REQUIREMENTS IN JAPAN

The usage of wheat in Japan is markedly different from that in the United States or other Western countries. Bread in the United States is the major outlet for wheat production. The JFA influences usage in Japan through its purchases of all wheat from both domestic production and international suppliers. Total wheat consumption in 1987 was approximately 6 million metric tons, including feed. Wheat for food accounted for 80 percent, with 1.18 million metric tons going for bread and rolls, 1.39 million metric tons for noodles, and 2.35 million metric tons for cakes, biscuits, and crackers (Uchida, 1988). Bread has become increasingly important in Japan, but

FIGURE 10.3. Annual Average Household Noodle Expenditure, Price, and Quantity in Japan, 1965-1987

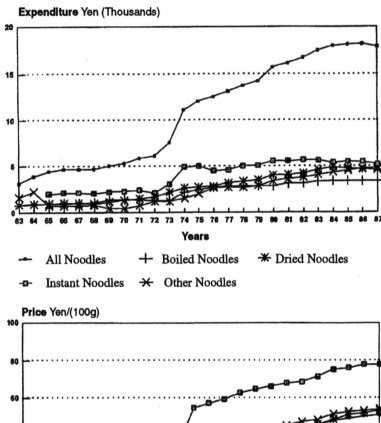

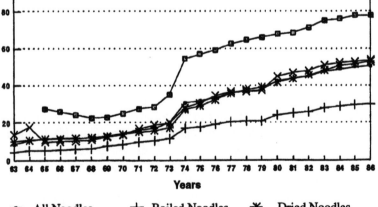

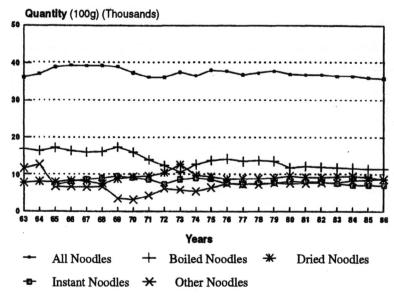

Quantity (100g) (Thousands)

━ All Noodles	┼ Boiled Noodles ✳ Dried Noodles
▫ Instant Noodles	✻ Other Noodles

Japanese Statistics Bureau. Family Income Survey.

both noodle and confectionery manufacturing are comparable users of wheat (Table 10.1). As a result, the secondary processing quality of Soft White Wheat from the Pacific Northwest is directed toward the peculiarities of confectioneries, baked goods (pastries, etc.), and certain types of noodles (Japanese noodles) consumed in Japan. The same is also true to a lesser degree for harder, higher-protein wheats (e.g., Hard Red Winter, Hard White Winter, Australian Standard White, and Canada Western Red Spring) from the United States, Canada, and Australia. These harder wheats are used because of their higher protein content, stronger gluten, and superior endosperm characteristics, which are important in both bread and noodle manufacturing.

Wheat manufacturing and processing in Japan exhibits varying degrees of concentration at each marketing level (Table 10.2). Flour milling is the most concentrated, with four firms accounting for almost 80 percent of the market.

The increase in bread consumption up to 1981 has been an important source for wheat demand in Japan. With the peak in bread

FIGURE 10.4. Annual Average Household Expenditure for Cakes and Confections and for Specific Cake and Biscuit Types, 1965-1987

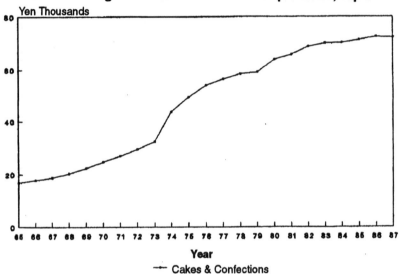

Annual Average Household Cakes Total Expenditure, Japan

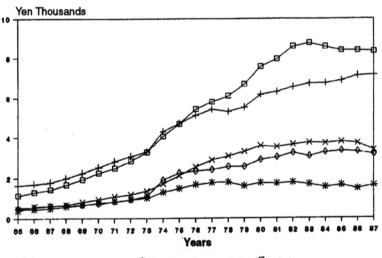

Japanese Statistics Bureau. Family Income Survey.

TABLE 10.1. Wheat Flour Production by Flour Usage in Japan, Selected Years, 1955-1987[1]

Year	Bread	Noodle	Pastry	Industrial	Family	Other	Total
			1,000 Metric Tons				
1955	668	874	277	—	262	—	2,081
1960	772	1,022	316	38		177	2,371
1965	997	1,167	403	98	112	201	2,977
1970	1,104	1,304	464	104	137	238	3,402
1975	1,410	1,449	559	120	176	282	3,996
1980	1,525	1,383	596	164	185	331	4,184
1982	1,609	1,466	605	161	190	338	4,369
1983	1,589	1,469	601	147	199	352	4,356
1984	1,599	1,543	590	141	193	380	4,445
1985	1,597	1,557	570	132	187	382	4,425
1986	1,626	1,597	590	119	192	400	4,524
1987	1,649	1,592	566	112	197	383	4,499

consumption in 1981, the highly automated bakery industry has had to take major steps to counter declining wheat consumption (Uchida, 1988). Strategies for improving product palatability, characterization, and differentiation are causing a restructuring of the bakery industry. Bakeries have continued producing bread in large automated lines but have chosen to rotate production schedules to also produce multiple confectionery and baked items in small amounts. Improved freshness and quality are also achieved by the preparation of partially frozen products and doughs, which undergo final baking at retail locations. This increase in the number of baked products places an even greater importance on flour quality. Wheat segregated by quality characteristics makes it possible to blend a uniform baking mix to meet the stringent specifications of baking technology (Pomeranz, 1987). If flour blends change, then it is very costly to both flour millers and bakers to make adjustments to flour blends or manufacturing processes.

Noodle manufacturing in Japan is less concentrated. Manufactur-

TABLE 10.2. Wheat Manufacturing: Industry-Level Concentration in Japan, 1989

CONCENTRATION

Industry	Number of Firms	Firm or Aggregation	Market Share
Milling	168	Nisshin	35%
		Nippon	22%
		Showa	8-10%
		Nitto	3-4%
		Share of Largest 10	79%
Baking	5,600	Largest 2	39%
		Yamazaki only	25%
		Middle size and Small	40%
		Baking Industry	60%
Wet Noodle	5,800	—	—
Dry Noodle	700	—	—

SOURCE: U.S. Wheat Associates, Japan, May, 1989.

ers purchase wheat flour from wholesalers or flour millers, so there is less interaction regarding quality (Table 10.2). Therefore, noodle manufacturers are very dependent upon flour millers to produce quality flour for the various noodle products they produce. Medium gluten strength and protein, clear-white or clear-yellow endosperm, and freedom from bran particles are important characteristics that determine the acceptability of the final noodle products. As the final noodle products vary in protein content, noodle manufacturing relies on flour millers to achieve the proper blend of wheat classes to produce each flour classification (Table 10.3). Japan extracts one to four grades of flour, with each noodle product utilizing different ratios of wheat class flours. The blending is highly dependent upon what is available or has been purchased by the JFA (Tables 10.3 and 10.4).

TABLE 10.3. Noodle Types and Noodle Flour Specifications in Japan

Type	Japanese Name	Moisture[1] %	Ash %	Protein %
Japanese Noodle:				
Very Thin Noodle	Somen	14.0	.38-.42	8.0- 9.5
Thin Noodle	Hiya-mughi	14.0	.36-.40	7.5- 9.0
Noodle	Udon	14.0	.40-.50	8.0-10.0
Flat Noodle	Himokawa	14.0	.38-.45	8.5-12.5
	Kishimen			
Binding Flour	Soba	13.5	.60-.95	9.0-12.5
Buckwheat	Tsunaghi-Ko			
Chinese Noodle:				
	Shina-Soba			
	Chuka-men			
Top Quality		14.5	.36-.38	9.5-12.0
Normal Grade		14.5	.38-.45	9.0-11.5
Instant		14.5	.40-.50	9.5-12.0

SOURCE: Mr. Simidsu, Technical Director, Milling Industry Development Foundation Communication with Western Wheat Quality Laboratory, Washington State University, 1970.
[1]These are approximate moisture percentages.

CHANGE IN WHEAT-CLASS PURCHASES BASED ON BEST END-USE CHARACTERISTICS

In Japan, price subsidies to producers and state trading are the most restrictive policies affecting wheat imports. Government (JFA) purchases and sales of imported wheat are expected to cover deficits incurred in government purchases and sales of domestically produced wheat. The JFA sells imported wheat at more than 50 percent above the purchase price. Intervention by the Japanese government has been shown to benefit producers, but it has been costly to consumers in terms of higher food prices and misallocation of resources. Because government intervenes in buying, price signals to producers (both domestic and international) do not convey

TABLE 10.4. Typical Blends of Wheat for Specific Noodle Flours in Japan

Noodle Type	Wheat Class or County Origin	Flour Ratio
Japanese Noodle:		
Thin Noodle	Japanese Domestic Wheat (SRW)	80
(Hand made)	Hard Winter	10
	Western White	10
Thin Noodle	Japanese Domestic Wheat (SRW)	30
	Western White	30
	Australian FAQ	40
Noodle (wet)	Japanese Domestic Wheat (SRW)	30
	Western White	30
	Hard Winter	20
	Australian FAQ	20
Noodle (boiled)	Japanese Domestic Wheat (SRW)	20
	Australian FAQ	50
	Hard winter	30
Noodle (boiled)	Japanese Domestic Wheat (SRW)	20
	Western White	45
	Australian FAQ	35
Noodle (dry)	Japanese Domestic Wheat (SRW)	20
	Western White	55
	Australian FAQ	25
Chinese Noodle:	Australia, Semihard, or Hard Winter	100
	Australia, Semihard, or Hard Winter	85
	Dark Hard Winter or Hard Winter	15
	Soft White	50
	Dark Hard Winter	50
	Western White	20
	Hard Winter	45
	Dark Hard Winter	20
	Dark Northern Spring or Manitoba (#2,#3)	15
	Western White	30
	Hard Winter	30
	Dark Hard Winter	40
	Western White	20
	Australian FAQ	30
	Hard Winter	20
	Dark Northern Spring or Manitoba (#2,#3)	30

Note: White Club Wheat is not acceptable for Chinese noodles.
SOURCE: Brief Information on Japanese Noodle, Hiroaki Shimidsu. 1970.
  Communication with Western Wheat Quality Laboratory, Washing-
  ton State University.
FAQ = Fair Average Quality.

manufacturing requirements. Some food processors in Japan have been forced to locate overseas. Elimination of the JFA role in wheat purchasing under trade negotiations of the Uruguay round would restructure the wheat market in Japan.

While the suggestion of eliminating the JFA seems radical, the same was thought to be true for the Livestock Industry Promotion Corporation (LIPC). The LIPC, a state trading organization responsible for administering policies for the livestock industries in Japan, was targeted for phased elimination in 1989. The JFA and the LIPC performed comparable functions in regulating imports and prices to processors and final consumers, as well as in determining price supports to producers.

The combination of wheat-class purchases in Japan for different flour strengths is currently dictated by the JFA. Political influences—as well as the necessity of combining wheat classes to produce desired end-product quality attributes for wheat products—are important considerations for foreign wheat purchases (Figure 10.5).

Nagao et al. (1979) tabulated the relative strength of flour derived from different wheat classes. Utilizing this information, and 1986 imports of wheat by class to Japan, we have estimated current demand for wheat flour by wheat class for each wheat-flour strength and each product that can be made (Figure 10.6).

Under present political considerations, with JFA purchasing from all sources, the semi-strong and medium flour market is shared between major suppliers. These two flour-market segments are the most likely to be affected by any relaxation of wheat trade restrictions. This market is presently estimated at 756,0000 metric tons of flour for Chinese noodles (semi-strong flour), and at 816,600 metric tons for Japanese noodles and all-purpose flour (medium flour).

U.S. Hard Red Winter, the most-produced U.S. bread wheat, would be the wheat class least preferred in the semi-strong and medium flour markets. Canada Western Red Spring is superior to Hard Red Winter for bread manufacturing, and Australian Standard White is preferred to Hard Red Winter for noodle manufacturing (Dick et al., 1986). Japanese Soft Red Winter is the least preferred wheat class for Japanese noodles, compared with Australian Standard White and Western White (Nagao and Sato, 1989). In addition, the Japanese Soft Red Winter industry is highly dependent upon

FIGURE 10.5. Flour Classifications and Types of Wheat in Japan

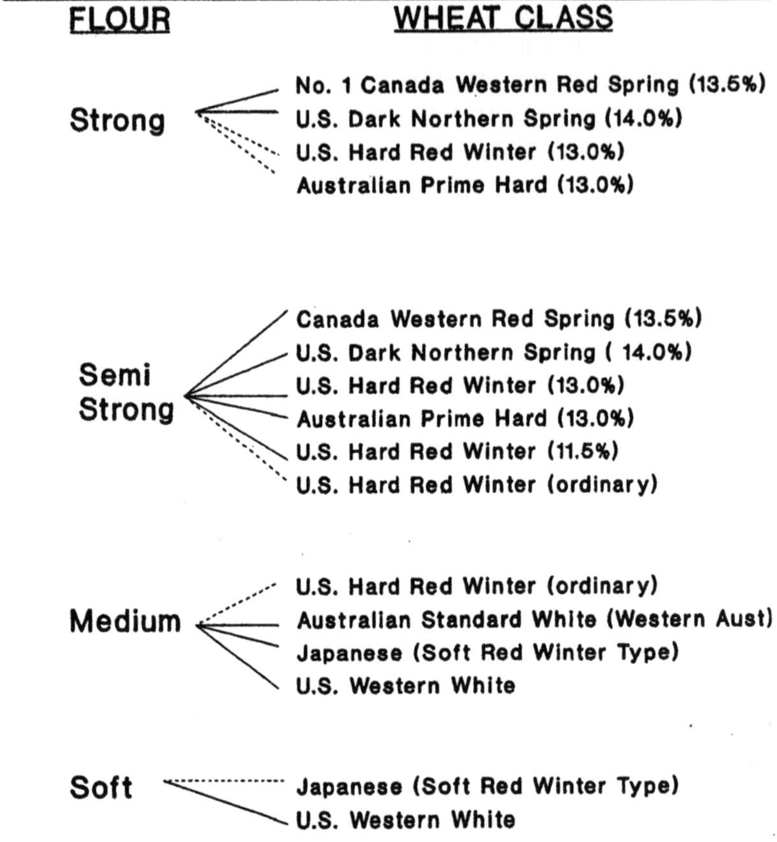

FLOUR   WHEAT CLASS

Strong

    No. 1 Canada Western Red Spring (13.5%)
    U.S. Dark Northern Spring (14.0%)
    U.S. Hard Red Winter (13.0%)
    Australian Prime Hard (13.0%)

Semi
Strong

    Canada Western Red Spring (13.5%)
    U.S. Dark Northern Spring (14.0%)
    U.S. Hard Red Winter (13.0%)
    Australian Prime Hard (13.0%)
    U.S. Hard Red Winter (11.5%)
    U.S. Hard Red Winter (ordinary)

Medium

    U.S. Hard Red Winter (ordinary)
    Australian Standard White (Western Aust)
    Japanese (Soft Red Winter Type)
    U.S. Western White

Soft

    Japanese (Soft Red Winter Type)
    U.S. Western White

Note: Solid lines indicate the most commonly used wheat classes for each flour classification and dashed lines less commonly used wheats.
SOURCE: Nagao, S. *Cereal Foods World,* Vol. 24, No. 12, December 1979, p. 593-95.

FIGURE 10.6. Estimates of Current Wheat Class Volumes Used in Production of Specific Flour Classifications and End Products in Japan, 1986.

Classification	Wheat Product	CWRS	U.S. DNS	U.S. HRW	Australian Prime Hard	ASW	Japanese SRW	U.S. WW	TOTAL
				1,000 M/T Flour					
Strong	Bread	391.0	344.6	291.0	107.1				1,133.7
Semi Strong	Bread	130.4	114.9	97.0	35.7				378.0
	Chinese Noodle	130.4	114.9	97.0	35.7				378.0
	Instant Chinese Noodle	130.4	114.9	97.0	35.7				378.0
Medium	Japanese Noodle			97.0		166.6	121.8	22.9	408.3
	All Purpose			97.0		166.6	121.8	22.9	408.3
	Confections			97.0		166.6	121.8	22.9	408.3
Soft	Confections						365.3	620.3	985.6
Total Wheat Class Use		782.2	689.3	873.0	214.2	499.8	730.7	689.0	4,478.2

ASSUMPTIONS

1. Thirty percent loss for processing bulk wheat to flour utilizing 1986 Japanese Wheat Import data.
2. Australian Wheat imports estimated to be approximately 30% Australian Prime Hard and 70% Australian Standard White.
3. Wheat Flour production as reported to JFA by Flour Millers: Bread flour 36.2%, Japanese Noodle flour 15%, Chinese Instant Noodles 7%, Chinese Noodles 8%, Pastry Flour 13%, and All Purpose 4%.
4. Wheat Classes are used for a flour strength classification are equally allocated between two strengths and then equally allocated to a wheat product. Exceptions to this procedure are ASW which is allocated 100% to noodle production and Western White which is allocated 10% to medium flour and 90% to soft flour.
5. Medium and soft flour are higher than expected since some flour is used for products not estimated.

109

government price supports to producers. Whether Japanese producers would continue to produce Soft Red Winter without price supports is unlikely, since resources such as land would probably be
channeled toward more competitive uses.

If U.S. wheat producers were to produce Hard White Winter in
volumes large enough to attract an international export market, then
the market currently satisfied by Australian Standard White, Australian Prime Hard, Japanese Soft Red Winter, and U.S. Hard Red
Winter (i.e., the semi-strong and medium flour for noodle manufacturing and all-purpose) would be the arena of competition. This
market represents 2.36 million metric tons of (bread, noodle, and·
all-purpose) flour as estimated (Figure 10.6).

We estimate the potential free-market trade for U.S.-produced
Hard White Winter at 412,800 metric tons of flour (Figure 10.7).
This potential Hard White Winter market represents only 18 percent
of the semi-strong and medium flour markets, assuming U.S. Hard
White Winter displaces U.S. Hard Red Winter and Japanese Soft
Red Winter based upon end-use preferences for noodle production.
The unshaded and shaded boxes in Figure 10.7 depict the end-use
categories that would change (positively and negatively) if U.S.
Hard White Winter were not produced and all Japanese trade sanc-
tions were lifted. Australia, as a wheat supplier to Japan, would
substantially gain market share in the semi-strong and medium flour
markets. The agronomic outcome of Hard White Winter produced
in the U.S. will determine its eventual market share and how well it
will compete with other wheat classes for these two flour markets.

Other wheat classes (e.g., Canada Western Red Spring, Western
White, Australian Prime Hard, and Australian Standard White)
would gain market share under free-market conditions if (1) there
were no Hard White Winter production (Figure 10.7) and (2) pro-·
tein content for Western White does not trend upward. In the free-
market scenario, with JFA no longer dictating wheat market purchases, U.S. Hard Red Winter is the wheat class most likely to lose
market share in the semi-strong to medium flour segment, regardless of whether Hard White Winter is produced or not. (Hard Red
Winter represents less than 9 percent of the 1988 Washington wheat
crop [Washington Agricultural Statistics, 1989]).

FIGURE 10.7. Free-Market Estimates of Wheat-Class Volumes Used in Production of Specific Flour Classifications and End Products in Japan, 1986

Classification	Wheat Product	CWRS	U.S. DNS	U.S. HRW	Australian Prime Hard	ASW[2]	Japanese SRW	U.S. WW	TOTAL	U.S. HWW[3] Potential
				1,000 M/T Flour						
Strong	Bread	391.0	344.6	291.0	107.1				1,133.7	
Semi Strong	Bread	227.4	114.9		35.7				378.0	
	Chinese Noodle	130.4	114.9		35.7	97.0			378.0	97.0
	Instant Chinese Noodle	130.4	114.9	0	35.7	97.0			378.0	97.0
Medium	Japanese Noodle			0		385.4	0	22.9	408.3	97.0
	All Purpose			97.0		288.4	0	22.9	408.3	121.8
	Confections			97.0		166.6	0	144.7	408.3	
Soft	Confections						0	985.6	985.6	
Estimated Free Market Trade by Wheat Class		879.2	689.3	485.0	214.2	1,034.4	0	1,176.1	4,478.2	412.8

Note: 1. Assumptions for Figure 6 hold for initial estimates of Figure 7.
2. Best end use purchasing shifts from U.S. HRW and Japanese SRW and more than doubles ASW demand when no U.S. HWW is produced.
3. Potential U.S. HWW production prevents a total best end use shift to ASW by 77% thus keeping lost U.S. HRW share and capturing some of the Japanese SRW share.

Increasing and | decreasing end product category of wheat class consumption under free market conditions

111

CONCLUSION

Japan's agricultural policies and agricultural sector are faced with possible reform as a result of the Uruguay round of the GATT multilateral trade negotiations which began in 1986. These negotiations will have repercussions in changing the structure of the wheat industry and, potentially, the way wheat is purchased in Japan. At the same time, wheat demand for bread, noodles, and bakery products has reached a plateau. Manufacturing and processing techniques are attempting to counter this trend by trying to preserve palatability, freshness, and taste and by diversifying products. With these types of changes comes an increased quality consciousness on the part of both manufacturers and consumers. Inherent quality of wheat purchased from exporting countries is the main input to quality in manufacturing.

The competitiveness between wheat classes for specific end uses depends upon the production, handling, and grading practices of exporting countries. Also very important is the selection of wheat varieties released with regard to characteristics important to both primary (bread) and secondary (noodle and baking) manufacturing sectors in Japan. Washington State, as the U.S.' largest producer of Soft White Wheat for export, primarily to Pacific Rim countries, needs to consider the relative importance of baking and noodle-manufacturing sectors. In order to maintain its market share of the confectionery all-purpose flour market in Japan, Washington State needs to protect those quality characteristics essential to the demands of that product market–mainly, low-protein and soft-gluten structure. Low protein and soft gluten in exported Western White are vital to the baking quality of confections and are also important in Japanese-noodle flour blends for Japanese manufacturing.

Marketing strategies and developments for wheat from Washington need to protect the important end-use characteristics of the Soft White Wheat market, namely low protein. By our estimate, the changing political climate may even increase this wheat class market if Japanese Soft Red Winter production declines and U.S. Soft White Wheats do not suffer increased protein levels.

The introduction of Hard White Winter wheat production in the U.S. is one strategy for competing with Australia in the Japanese wheat market. In light of the present findings, it is apparent that Hard White Winter would also compete with U.S.-produced Hard Red Winter for noodle flour. The Hard Red Winter market will decline even without Hard White Winter production, if the political influences on Japanese wheat purchasing are relaxed and buying is then based upon best end-use properties of each wheat class.

For Washington State, Hard Red Winter production is not important in the wheat export market with Japan. However, Hard White Winter production has important implications for Washington Soft White export value if utmost care is not taken in preventing blending of the two wheat classes. Lack of visible wheat-class distinction between Hard White Winter and Soft White Winter, along with the practice of blending, would dilute the end-use characteristics most important to Japanese confectionery production, namely low protein and soft gluten. The potential Hard White Winter market, as estimated, represents approximately 27 percent of the total wheat-flour market in Japan. It is unlikely Japan will totally substitute Hard White Winter for Australian Standard White and other wheat classes, unless it is of a superior variety for noodle manufacturing. However, Hard White Winter production could prevent Australian Standard White from doubling its market share at the expense of the lost Hard Red Winter market.

Chapter 11

Japanese Preference for Bakery Food Ingredients

Yeshajahu Pomeranz

This review covers the following areas of Japanese preference for bakery food ingredients: (1) the usage of soft white wheat in production of sponge cake and Oriental noodles; (2) requirements for bread wheats and flours (including the effects of milling, dough mixing, and class differences); (3) recent developments in the use of improvers (oxidants, gluten improvers, enzymatic preparations, frozen and refrigerated doughs, antistaling agents); (4) shortenings; and (5) consumer trends in food purchases.

In Japan, staple food (rice, wheat, and barley) transactions are controlled by the government under the Food Control Act (Sawabe and Uchida, 1978). Japanese flour mills, therefore, are not allowed to import and select the types of wheat that they wish to use. The government is responsible for all imports of wheat for food and feed use, and it releases the wheat to the flour mills.

There are over 150 flour-milling companies in Japan, most of which are general flour millers. About half are general flour millers as well as Zosan millers (Increased Bran millers) and about two dozen are Senkan millers (Special Bran millers).

Japan imports about 5.5 million tons of wheat per year, and only about 0.2 million tons of domestically grown wheat is delivered to the government. Japanese mills depend on wheats from U.S., Canadian, and Australian sources. The mills must use the available domestically grown wheat to be eligible to purchase imported wheat from the Japanese Food Agency.

General flour mills take off about 78 percent flour, but the Zosan and the Senkan mills are limited to a maximum of 45 percent flour extraction. Japan's milling operation is much the same as in the rest of the world, varying only in the Food Control System and in the high usage of flour by the Oriental noodle industries.

Wheat processing and consumption trends in Japan were reviewed by Nagao (1979). Out of a total of over 4 million tons of flour, about one-third each was used to make bread and noodles; about 15 percent for confectionery; and the rest for family, industrial, and other uses.

Originally, since the end of World War II and up to the late 1970s, much of the concern in the Japanese flour-milling and baking industries has been with the highly diverse identity, nature, and functional properties of flours milled from wheats imported from many countries. Those were summarized by Nagao et al. (1976, 1977) in two papers dealing with quality characteristics of soft wheats and their use in Japan.

SOFT WHITE WHEAT

Since usage of soft wheats in the Japanese market is different than that in Western countries, quality requirements for them have unique aspects (Nagao et al., 1976). Sponge-cake and Japanese-type noodle testing methods were found most valuable in evaluating the secondary processing quality of soft wheats. General quality evaluation of soft wheats can be derived from the combination of the results of sponge-cake and Japanese-type noodle tests, as well as by the American Association of Cereal Chemists (AACC) cookie test and by consideration of test milling and analytical results.

In a subsequent study (Nagao et al., 1977), soft wheats–such as Soft White, White Club, and Soft Red Winter wheats from the U.S.–Victoria Soft, Victoria F.A.Q., Western Australia F.A.Q., French, and domestic Japanese wheats were compared with respect to their utility for Japanese products. White Club, Soft White, and Soft Red Winter wheats were superior to others in suitability for confectionery products, although White Club was considered better than Soft White. Small quantities of Victoria Soft and some types of French and Japanese wheats can be blended with soft wheats from the U.S. in the production of confectionery flour. In spite of the low protein content, the kernel characteristics of both Australian F.A.Q. wheats were rather hard, and they were least preferable in terms of their sponge-cake baking qualities. As material for Japanese-noodle flour, wheats similar to Japanese wheat were considered most desirable. Although Australia F.A.Q., Soft White, and White Club wheats were different from Japanese wheat, they possessed favorable characteristics for noodle flour.

The perception of the superiority of Australian Soft White wheats in the production of Japanese noodles has been continued in many subsequent publications from Japan and the U.S. In the first of two studies conducted by Toyokawa et al. (1989a), four wheat flours (a commercial Japanese-noodle flour, a Soft White and a Club wheat from the Pacific Northwest, and an Australian Standard White wheat) varying in noodle-making quality were used to investigate the role of flour components. A fractionation and reconstitution interchange of gluten, primary starch, tailing starch, and water solubles was used to investigate the role of each in udon noodle quality. The primary and tailing starch fractions were most responsible for noodle texture. Of the two, the primary starch fraction contributed the most to the desirable viscoelasticity of noodle texture.

In the second paper (Toyokawa et al., 1989b), the physical and chemical characteristics of primary and tailing starch fractions from four wheats that differed in noodle quality were studied. The particle-size distribution, water-holding capacity, and amylose/amylopectin ratio (iodine blue color) were determined. The most apparent difference was the water-holding capacity at $75°C$ and the amylose content. The water-holding capacity of these flours was highly correlated with the viscoelastic textural properties of noodles. Increased levels of amylose decreased the water binding of cooked noodles and increased the associated firming and loss of elasticity. Using pure corn starches differing in amylose/amylopectin ratio as a replacement for the inherent primary starches substantiated the relationship between amylose content and noodle viscoelastic properties. The viscoelastic properties of noodles appeared strongly associated ($r = 0.85$) with the thermophysical properties of the starch, as measured by water-holding capacity of the starch at $75°C$. There was a highly negative relationship ($r = -0.96$) between amylose content and water-holding capacity at $75°C$.

The role of flour constituents of soft wheats in Japanese sponge cake was studied by Noguchi and Rubenthaler (1978). As stated before, soft white wheat from the U.S. Pacific Northwest area is the most important wheat for confectioneries in the Japanese market. For soft-wheat quality evaluation, a Japanese sponge-cake baking method was most useful. The components of wheat flour that are most responsible for sponge-cake baking quality were examined–as in the previous studies–by using a fractionation and reconstitution technique. Composite flour samples from Club and common Soft Spring were fractionated into gluten, starch, and water solubles, interchanged one fraction at time, reconstituted, and baked. The gluten fraction was responsible for baking quality. The starch fraction had a slight effect, while the water soluble fraction had little effect on baking quality. All the baking quality difference between the two subclasses of wheat could not be explained by merely interchanging the fractions studied.

By baking the reconstituted flours at different protein levels (from 6 to 12 percent), it was found that protein content was significantly and inversely correlated to baking quality. Such correlation was also confirmed by baking several samples of different varieties, locations, and crop years in the Pacific Northwest. The correlation between cake score and other quality factors–such as water absorption, alkaline water-retention capacity, viscosity, and cookie diameter–was also examined. The correlation coefficient between cake score and viscosity was significantly high ($r = -0.77$).

More recent trends in uses of soft wheats were described by Nagao (1988). They were summarized as follows: U.S. Western White, Australian Standard White, and Japanese wheats are used for milling soft wheat flours that may be divided into three general categories–low-protein flours of various extractions for confections, Japanese-noodle flours, and all-purpose flours (including family use). For the production of certain confectionery flours, a special- or top-grade cake flour is extracted from high-quality soft wheat, of which the most important is U.S. Western White wheat. A low protein content, desirable softness of gluten quality, and low α-amylase activity are the primary requirements in a flour of this type. The superiority of Australian Standard White wheat to others for the production of Japanese-noodle flours was proved by a noodle test, although the wheat is usually blended with Japanese and/or U.S. Western White wheats in the commercial milling. Gelatinization properties of Australian Standard White and Japanese wheats were compared with respect to their suitability for Japanese noodles.

Flour quality requirements of staple foods in Asia and the Middle East were reviewed by McMaster and Moss (1989). Pan bread requires a flour that has a moderate amount of starch damage, a protein of at least 11 percent, high dough stability, and intermediate paste viscosity. At the other extreme are biscuit and cake flours, which should be low in starch damage, protein, and dough stability. Noodles have intermediate requirements for protein and dough stability, should be low in starch damage, and have a fairly high hot-paste viscosity.

BREAD WHEATS AND FLOURS

Insofar as bread-making quality of wheats is concerned, several factors/criteria are of interest and concern to the milling and baking industry. They include milling characteristics; dough strength and stability during mixing; the functional differences between Hard Red Winter and Hard Red Spring wheats; the need to store wheat (after harvest) and flour (after milling) before use; the effects of overgrinding; oxidation requirements; and total absence of pre-harvest sprouted grain.

In the review by Nagao (1979), wheats and flours produced from wheats imported into Japan were classified into strong (Canada Western Red Spring wheat, protein 13.5 percent, and U.S. Dark Northern Spring wheat, protein 14 percent, for bread production); semi-strong (U.S. Hard Red Winter and Australian Prime Hard wheat, protein 13 percent, for some bread types and Chinese noodles); medium (lower-protein U.S. Hard Red Winter and Australian Standard White for Japanese udon noodles and some types of confectionery); and soft (low-protein U.S. Soft White for confectionery). The flour-milling industry provides over 50 types of flour for various bread types, Chinese and Japanese-style noodles, and confectionery (Nagao, 1989). Tests of baking methods and their application in evaluating wheat flour in mills and in evaluating various ingredients in bakeries and allied companies were described by Uchida (1982). The tests included bread baking and Chinese-noodle making for medium- to high-protein wheat flours, as well as sponge cake, cookie, and Japanese-type noodle-making tests for soft, low-protein wheat flours.

The concern about the effects of new milling methods and high-speed dough mixers on bread quality is exemplified by a series of studies on flour overgrinding and dough overmixing. Overgrinding of hard-wheat flour streams affected starch damage, rheological properties, and oxidation requirements (Okada, Negishi, and Nagao, 1984). Factors affecting the rheological properties of doughs in terms of dough development were studied using a Brabender Do-Corder (Endo, Tanaka, and Nagao, 1983; Endo, Okada, and Nagao, 1985; Nagao, 1985). Okada (1989) studied overall factors affecting dough mixing; specific factors that are related to, or govern, dough breakdown during overmixing were researched by Okada, Negishi, and Nagao (1986). The above were related both to oxidation requirements and to flour strength, as governed by wheat- and flour-processing parameters.

A comparison of milling and analytical characteristics of Hard Red Spring and Hard Red Winter wheats was the subject of a study by Endo et al. (1988b). Reversed-phase, high-performance liquid chromatography (RP-HPLC) was used to analyze each of the 15 samples of Hard Red Spring (HRS) and Hard Red Winter (HRW) wheat. Wheat protein content of the commercial U.S. samples ranged from 12 to 14.4 percent. A difference was noted between the two wheat classes in the 70 percent ethanol-soluble protein patterns obtained from RP-HPLC. Two peaks with longer retention times were larger for the HRS than for the HRW wheats. Milling and flour analytical properties were compared with a Buhler laboratory mill. The total amount of break flour derived from the HRS wheats was less than that obtained with the HRW wheats. The reverse was noted for the reduction flours. Total flour yield for the HRS wheats was higher than for the HRW wheat. Somewhat higher values for damaged starch and ratios of starch tailings to total starch isolated were observed for the HRS wheat. There was little difference in the ratio of free lipid to total lipid content between the two wheat classes.

While conflicting results have been reported on differences in milling yields of wheats from the two wheat classes, there have been consistent claims about the better bread-making potential of HRS wheats, when compared with HRW, on an equal protein basis. It would seem that while, on the average, HRS wheats are superior to HRW wheats in bread-making potential, there is much variation among wheat varieties in both classes. However, regardless of the scientific and technical evidence (both from the standpoint of competition with other suppliers of HRS and the perception in the grain trade of its uniqueness), it is mandatory to provide our buyers with the preferred wheat type.

The preference for HRS was emphasized again by Nagao (1989). In that same study, Nagao also emphasized the superiority in noodle-processing characteristics of Australian Standard White (ASW) wheat to that of other soft wheats. This is thought to be related mainly to its starch characteristics. Amylograph gelatinization temperature for ASW flour is lower than that of other soft wheats. The relatively low gelatinization temperature was presumed to be a factor in soft and pliable noodles.

Gelatinization properties of ASW and wheats representing the three major varieties grown in Japan were compared with respect to their suitability for Japanese noodles (Endo et al., 1988a). The addition of sodium chloride caused a pronounced difference in the area under the sticking-out portion of the Brabender amylogram during the cooling stage for flours milled from ASW and Japanese wheats. ASW having favorable characteristics for noodle making showed a large area, whereas small or no area was observed for Japanese wheats. This may relate to noodle quality, especially to noodle staling.

IMPROVERS

It is impossible to describe in this short space the thousands of improvers, improver modifications, and improver combinations developed and used by the baking industry in Japan. Only a few major examples will be given.

Oxidants. Potassium bromate is the oldest and most widely used oxidant in bread-making. It is used to improve dough-handling properties, crumb grain and texture, and to increase loaf volume and shelf life.

The concern about the use of oxidants, in general, and of potassium bromate, in particular, was voiced by Uchida in 1979. Several aspects of the problem have been raised. They include the inherent oxidation requirements of flours milled from various wheat classes and varieties as well as the need to store (mature) freshly harvested wheat and freshly milled flour. This becomes of special significance in modern pneumatic milling, given the pressure to store minimum amounts of wheat in the mill and flour in the bakery.

Recent toxicological studies by Japanese researchers have shown that bromate is a genotoxic carcinogen. Previous work has shown no similar toxicological effects in rats fed bread or flour containing the maximum bromate level of 75 parts per million. The level used in practice is 20 to 40 parts per million; higher levels damage bread-making properties. A recent Japanese study of a bread baked with 75 parts per million bromate showed bromate levels at the parts-per-billion level.

In October 1989, the British government removed potassium bromate from the list of permitted flour improvers. Bromate is no longer permitted in Japan, and its addition has been prohibited for quite some time in most European countries. As a result of those studies, there is a large amount of research to find a "natural" replacement for potassium bromate in Japan. Some potential replacements include various lipoxygenase preparations.

Gluten improvers. The patents include some that are based on physical treatment (e.g., supercritical carbon dioxide); enzymic modification (e.g., controlled proteolysis); chemical modification (deamidation); interaction with modified lipids (e.g., lysolecithin from action of phospholipase A); and blends of gluten, lipase, and lecithin. A whole series of products is based on the use of sucrose esters–highly purified and varying over a wide range of hydrophilic-hydrophobic balance. It is of interest to note that the technology to produce lysolecithin (or the interaction product with gluten) is based on research originally conducted in Germany and the U.S. Sucrose esters are produced in Japan, based on patents developed in the U.S.

Enzymatic preparations. The patents include α-amylases of intermediate thermostability (to provide fermentable sugars and modify starch gelatinized during the baking process); amylolytic enzymes that can attack, to a small extent, intact starch granules; and lipases, lipoxygenases, oxidases, and proteases.

Frozen and refrigerated doughs. These are made possible through the use of special yeast strains or protectants to reduce freezing damage.

Antistaling agents. These act through starch modification. They are galacturomic acid-based, protein-based, and particularly mixtures of gums and glycerol fatty-acid esters.

SHORTENINGS

Japan produces an astounding and impressive array of shortenings, fillings, and spreads for the baking industry. They are specifically tailored for a great variety of products. A factory that produces under one roof–semi-automatically–up to 30 widely different types of baked products of the highest quality and uniformity is a most memorable sight. This is made possible by the use of additives and improvers (including shortenings).

Okamoto et al. (1989a, 1989b) reviewed fatty-acid compositions and physical characteristics of bakery margarines and shortenings on the market in Japan. The fatty-acid compositions and physical characteristics of bakery margarines were determined for five margarines for bread, cake, pie or pastry, puff pastry, and whipped cream (Okamoto et al., 1989a). Melting points were measured by four methods, with two types of automatic apparatus; solid fat content (SFC), the hardness index (by cone penetration), oil-off, and creaming values also were determined. Long-chain (C > 20) fatty acids were detected in relatively high proportions from 14 brands, indicating the addition of hydrogenated marine oil. From measurements of the above physical parameters, the margarines for pies or pastry were hardest, followed by those for puff pastry. Those for whipped cream were the softest. Softer types tended to oil-off more easily. The type for pies or pastry had a somewhat lower SFC (< 30°C) than the margarine for puff pastry did. The hardness index was correlated with SFC for bakery margarines.

Fatty-acid compounds and physical characteristics of shortening were determined for 25 brands of products: five brands each for bread, whipped cream, and frozen dessert; five fluid brands; four brands for cake; and one for pie crust (Okamoto et al., 1989b). Results of fatty-acid analyses indicated the main material oil or fat to be (1) a vegetable liquid oil in all brands of liquid shortening, (2) a laurin oil in two types for frozen dessert, and (3) a palm oil in the other three for frozen dessert. Hardened marine oils were blended in ten other types, as is evident from the relatively high levels of long-chain fatty acids (C > 20).

According to the SFC and hardness-index curves, the frozen desserts, particularly those containing mainly laurin oil, tended to be harder than other solid products at < 15°C, but they readily softened at higher temperatures. The SFC curve for fluid products indicated solid fat components to be comprised only of the emulsifiers added. The hardness index was correlated to SFC for the shortening. The total percent of unsaturated fatty acids, except trans-isomers, was also correlated to SFC at 20°C for all the products; if products comprised mainly of laurin oil were not considered, the correlation was even higher.

CONSUMER TRENDS IN FOOD PURCHASES

In common with most Western countries, there has been an increasing trend in Japan to produce more nutritious and more natural foods. The consumer is looking for convenience foods that can be prepared rapidly (without sacrificing quality) in microwave ovens. Quality expectations are actually rising. There is an increased concern about the caloric density of foods and especially of fats, in general, and saturated fats, in particular. The increase in consumption of refined foods and the health consequences associated with it (constipation, diverticular disease, colon cancer, diabetes, and arteriosclerosis) have prompted an increased amount of dietary fiber preparations. A plethora of such preparations is available; some are rich in soluble dietary fiber and some in insoluble dietary fiber. Many are available for incorporation into a variety of baked products and are exported to various countries in the Pacific Rim.

Chapter 12

Overview of Japanese
Fruit Consumption Trends
with Emphasis on Apple Demand

Manfred Heim

The Washington State apple industry has expended considerable
effort to gain access for its fresh apples in the Japanese market, but
current phytosanitary restrictions prohibit entry. Also, Washington
apple marketers face many unknowns even if granted clearance.
This paper gives an overview of the Japanese fruit market situation
and what may lie ahead for the Japanese apple market.

JAPANESE HOUSEHOLD CONSUMPTION TRENDS

Unlike U.S. produce consumption, Japanese purchases of fresh produce have declined. Over the 1970 through 1985 period, U.S. per
capita fresh fruit consumption increased by 12.4 percent, while Japanese data show a 15.5 percent decline in household consumption. This
decline occurred as Japanese fresh fruit prices (adjusted for inflation)
decreased by 2.4 percent and average real incomes increased by 45
percent. This raises the question of why fresh fruit purchases declined
in an environment of greater economic prosperity.

EXPENDITURE PATTERNS FOR FOOD IN JAPAN

The biggest gains in food expenditure from 1963 through 1986
went for food away from home, up 263 percent in real terms (see

125

Figure 12.1). Next came the "cooked foods" category, up 243 percent. Other large gainers included meat and seafood, up 109 and 88 percent, respectively, over the same period. Fruit and vegetable expenditures went up 53 and 51 percent, respectively. On the other hand, cereal grain expenditures declined by 13 percent over this same period.

JAPANESE APPLE INDUSTRY

Area and Production Trends

Among fruits in Japan, apples are second behind mandarin oranges, which have over twice the total area and production. Despite declining household apple consumption, the total area of apple production has remained relatively constant in the last two decades (see Tables 12.1 and 12.2). Some varieties, such as Red and Golden Delicious, have shown marked declines. Others, such as Fuji and Tsugaru, have increased. These last two varieties now account for over half the Japanese production; this contrasts with the Red and Golden Delicious varieties, which in the mid-1970s made up 40 percent of total production. However, by 1988, the Delicious varieties made up only 16 percent of the total production.

Costs and Returns

The average production costs per hectare for apples have increased over time but appear sensitive to yield of marketable product. From 1979 through 1986, the varieties that had the greatest production cost (Fuji, Mutsu, and Tsugaru) also provided growers the highest returns (see Table 12.3). Red and Golden Delicious cost relatively less to produce, but gave substantially lower returns. A shift from the Delicious varieties seems driven by higher yields, lower production costs, and overall greater returns for other varieties.

If Washington Red Delicious, for example, could substitute for Fuji, they would sell at much higher prices than in the U.S. market. At 1986 price levels, Japanese wholesale prices for Fuji reflect a

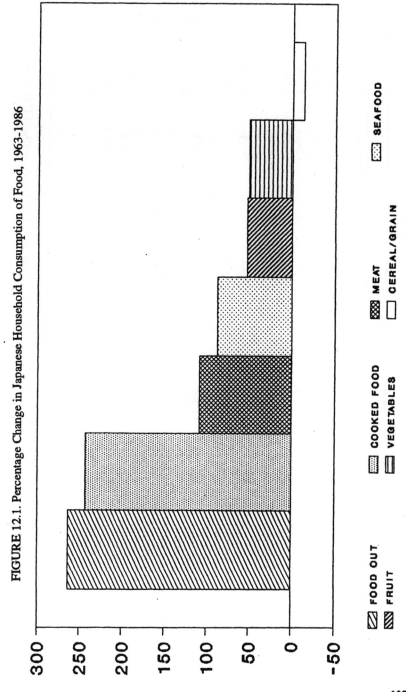

FIGURE 12.1. Percentage Change in Japanese Household Consumption of Food, 1963-1986

FOOD OUT   COOKED FOOD   MEAT   SEAFOOD

FRUIT   VEGETABLES   CEREAL/GRAIN

TABLE 12.1. Bearing and Non-Bearing Area by Apple Variety (Hectares) in Japan, 1970-1987

Year	Bearing Area by Variety							Total	Non-Bearing Area
	Red Delicious	Golden Delicious	Fuji	Tsugaru	Matsu	Jonathan	Other		
1970/71	10,900	3,580	0	0	0	11,400	30,220	56,100	3,500
1971/72	12,800	4,150	0	0	0	10,800	27,050	54,800	4,200
1972/73	14,200	4,190	0	0	0	9,310	25,600	53,300	4,900
1973/74	15,500	3,990	5,840	0	0	7,970	18,300	51,600	5,300
1974/75	15,800	3,700	7,670	0	0	6,690	16,240	50,100	4,900
1975/76	16,100	3,430	8,410	0	1,920	5,690	13,050	48,600	4,600
1976/77	16,700	3,060	9,710	0	1,980	5,030	11,020	47,500	3,900
1977/78	17,000	2,680	10,700	0	1,980	4,210	10,030	46,600	4,100
1978/79	17,000	2,430	11,900	0	1,940	3,890	9,740	46,900	3,800
1979/80	16,200	2,130	13,100	1,560	1,870	3,300	8,340	46,500	4,200
1980/81	15,400	1,900	14,300	2,180	1,870	2,860	7,890	46,400	4,800
1981/82	14,700	1,720	15,700	2,660	1,850	2,600	7,370	46,600	5,400
1982/83	13,900	1,590	16,900	3,190	1,820	2,380	5,220	45,000	8,100
1983/84	12,800	1,420	17,800	3,810	1,760	2,060	5,750	45,400	8,500
1984/85	11,800	1,280	18,800	4,300	1,710	1,860	5,850	45,600	8,700
1985/86	11,100	1,050	19,700	4,860	1,720	1,670	5,600	45,700	8,700
1986/87	N/A	N/A	N/A	N/A	N/A	N/A	N/A	46,200	8,500
1987/88	N/A	N/A	N/A	N/A	N/A	N/A	N/A	46,500	8,300

Source: U.S. Dept. of Agriculture, Foreign Agricultural Service. Foreign Production Estimates Division. Horticultural and Tropical Products Division. *World Apple Survey, a Statistical Compendium*, July, 1987, and unofficial USDA, FAS data on selected fruit items.

TABLE 12.2. Total Production (Metric Tons) by Apple Variety in Japan, 1970-1987.

Year	Red Delicious	Golden Delicious	Fuji	Tsugaru	Matsu	Jonathan	Other	Total
					Variety			
1970/71	183,000	65,300	0	0	0	218,200	554,200	1,021,000
1971/72	223,100	74,100	0	0	0	204,000	505,800	1,007,000
1972/73	264,300	74,800	0	0	0	169,200	450,700	959,000
1973/74	301,200	75,100	91,000	0	25,500	156,700	313,200	962,700
1974/75	273,900	60,800	123,100	0	26,500	112,900	253,200	850,400
1975/76	308,100	60,000	159,200	0	32,000	102,800	235,900	898,000
1976/77	319,400	55,400	185,500	0	32,500	94,200	192,400	879,400
1977/78	361,900	50,800	240,300	0	36,400	87,500	181,900	958,800
1978/79	278,700	42,500	259,800	0	31,200	72,100	159,700	844,000
1979/80	304,600	36,200	252,000	30,300	33,800	60,100	135,800	852,800
1980/81	334,000	35,700	317,400	43,000	37,000	56,800	136,200	960,100
1981/82	253,300	28,600	324,500	50,500	32,700	42,100	114,000	845,700
1982/83	282,400	26,500	354,800	59,000	34,500	41,300	125,000	923,500
1983/84	278,600	27,500	447,300	80,700	37,000	41,200	135,700	1,048,000
1984/85	185,900	17,500	359,700	76,600	28,500	27,600	115,900	811,700
1985/86	194,500	16,800	414,000	98,400	30,800	26,900	128,400	909,808
1986/87	170,300	15,800	464,600	126,700	34,400	25,900	140,700	986,100
1987/88	149,000	13,500	493,200	123,200	32,900	23,000	168,200	1,003,000

Source: (See Table 1).

TABLE 12.3. Japanese Average Apple Production Costs and Value, 1979-1986, and Average by Variety, 1986

Year and Variety	Total Cost (Yen/100 kg)	Labor Cost (Yen/100 kg)	Value (Yen/ha)	Estimated Cost (Yen/ha)	Yield[1] (kg/ha)	Grower Price (Yen/100 kg)
1979	14,044	7,091	487,645	339,865	24,200	20,151
1980	11,933	5,904	457,971	353,694	29,640	15,451
1981	14,081	6,925	518,625	370,894	26,340	19,690
1982	14,374	7,211	411,907	379,330	26,390	15,608
1983	13,731	7,015	377,908	393,118	28,630	13,200
1984	17,236	8,870	508,313	385,914	22,390	22,703
1985	15,393	7,468	517,875	394,391	25,620	20,214
1986	15,056	7,667	424,967	420,283	27,920	15,221
Variety 1986:						
Starking Delicious	12,986	6,079	185,404	304,676	23,460	7,903
Fuji	13,793	7,238	558,636	474,439	34,400	16,239
Tsugaru	14,711	7,158	539,525	504,258	34,280	15,739
Golden Delicious[2]	12,701	6,897	277,738	301,014	23,700	11,719
Mutsu	14,157	8,378	652,437	445,256	31,450	20,745

Source: Statistics and Information dept. Ministry of Agriculture, Forestry, and Fisheries. The 63rd Statistical Yearbook of Ministry of Agriculture, Forestry, and Fisheries Japan, 1986-87. March 1988.
1. Yield of marketable product.
2. 1984 data.

price of $20 per box to Washington F.O.B. This would attract exporters of Extra Fancy Delicious of preferred sizes. However, these price relationships do not reflect a reliable guide for the future. Future prices will depend on exchange rates and relative prices. Seasonal and varietal issues will also play a part.

Varietal and Seasonal Factors

The marked decline in market share for previously popular Delicious varieties reflects a preference for certain Japanese varieties. Japanese Red Delicious, produced in a humid coastal climate, do not have the same size, color, or taste of Washington Red Delicious. Washington Red Delicious would have to compete for a quality marketing niche distinct from Japanese Red Delicious. The main competition would come from varieties like Fuji. Fuji apples accounted for 30 percent share of the 1979 Japanese apple crop, compared with 36 percent for Starking Red Delicious. By the 1987-88 marketing season, Fuji's production share had increased to 49 percent, while Japanese Red Delicious share had declined to 14.9 percent.

Most Japanese apples are sold in the months of October through December (see Figure 12.2). Small quantities of storage apples are available for about six months after harvest, but Japan has nothing to compare with the scale of Washington's controlled atmosphere (CA) storage system. Japanese Red Delicious are also harvested at a mature stage, thereby reducing the opportunity to store these apples. Fuji, however, has an advantage under Japanese conditions because of a relatively longer storage life (extending into April or later).

The question for Washington exporters centers on whether to market to Japan continuously throughout the year or to focus on either early-, middle-, or late-season dates when Japanese production declines. Early-season exports, from October through December, would face the heaviest domestic competition. Mid-season, from January through March, would face reduced competition but also uncertainty as to how Japanese consumers would react to CA storage apples. From April through the summer, the stiffest competition may originate from Southern Hemisphere producers like Australia, New Zealand, and perhaps Chile.

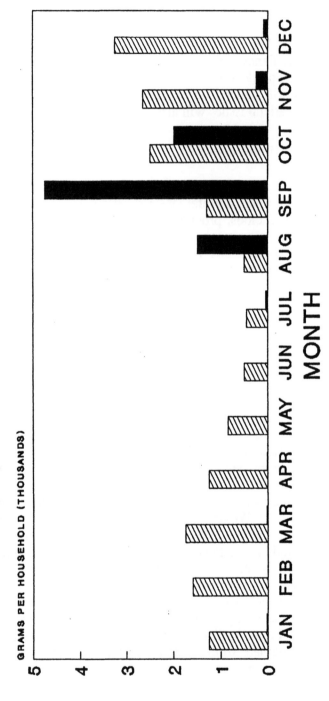

FIGURE 12.2. Quantity of Purchased (Grams) of Apples and Pears in Japan by Month, 1986

132

PROJECTING JAPANESE APPLE DEMAND

Apple consumption, as previously explained, has decreased as part of a general decline in overall household fresh fruit consumption–from a peak in the 1960s of about 7 kilograms per person to about 5 kilograms per person in 1972. Apple consumption over the last ten years has not declined to the extent it did during the late 1960s through the mid-1970s. From historical data, it appears that income growth has had little impact on household apple consumption. Real apple prices have, however, risen since 1962. Demand appears to be price-sensitive in accordance with economic theory; that is, an inverse relationship exists between quantity demanded and its own price. To quantify these relationships, we estimated a demand model for Japanese household consumption of apples.

Estimated Demand for Apples in Japan

We estimated a demand model for apples in Japan by using annual average-time series data from 1963 through 1986. Data were obtained from the *Annual Report on the Family Income and Expenditure Survey, 1986*, published by the Statistics Bureau of the Management and Coordination Agency of Japan. Equations were estimated in both a price-dependent and quantity-dependent form.

Retail Apple price = f (Apple quantity, Pear quantity, Income),

Apple quantity = f (Apple price, Pear price, Income),

where f means a function of.

All quantities and income were on a per capita basis, and price and income were deflated by the consumer price index.

The model in the price-dependent form assumes that average price in any year is affected by the given supplies and levels of income in that year. It permits an evaluation of possible price effects due to supply increases, either from Japanese sources or from imports. The model in the quantity-dependent form permits us to evaluate the effects of price and income on the annual average per capita consumption of apples. Thus, we can make a prediction of future consumption given certain assumptions pertaining to income growth and prices. Both models were estimated using ordinary least squares (OLS). This is appropriate if we assume that price does not simultaneously affect supplies. The results are reported in Table 12.4.

Both price-dependent and quantity-dependent models generated the expected relationships between apple quantity and its own price and between the substitute (pear) and apple price or quantity. The income effect was negative in both equations, but its overall impact within both equations was relatively small. The price elasticity measured at the means in the quantity-dependent model was −0.57 or inelastic, meaning a 1 percent change in price results in a less than 1 percent change in quantity consumed.

To project what may occur in the future in terms of possible price effects and consumption levels, we made certain assumptions. Holding all other variables constant (at their means) except income, we estimated what an income growth of 2 percent per year over the next ten years would mean in terms of per capita apple consumption for Japan in 1996. This resulted in an estimate close to 5 kilograms (11 pounds) per capita. This can be compared with average consumption over the last ten years, of 4.94 kilograms (10.9 pounds) per person. Both these figures are well below the historical average of 5.45 kilograms from 1963 through 1986. If we assume a very large annual increase in income of 7 percent, then predicted consumption declines to about 4.5 kilograms (10 pounds) per person. Clearly, it appears that rising income levels in Japan have not necessarily meant increased apple consumption.

The price-dependent equation can be used to predict how imports will affect the price for apples in the Japanese market. Assuming Japanese apple supplies are at the average level for 1963-86, then with no imports, we project apple prices in real terms to be 29.1 yen/100 grams, or a 0.94 yen/100 gram decline in real apple prices at the retail level. If consumers treated U.S. apples and Japanese apples as direct substitutes, one million boxes of U.S. apples imported into Japan would, in a typical year, lead to an average price decline of about 3 percent. However, there is reason to believe that Washington Red Delicious will not be direct substitutes for Japanese apples.

It should be noted that the two-stage least squares algorithm was

TABLE 12.4. Estimation of Japanese Apple Demand, 1963-1986.

Dependent Variable	Independent Variable				Standard Error	F.	R^2	DW
	Constant	Apple Quantity	Pear Quantity	Income				
Apple Price	72.39	–.600	–.373	–.0000071	1.93	30.13	.82	1.34
	(6.12)*	(–4.81)*	(–1.89)	(–.273)				
	Constant	Apple Price	Pear Price	Income				
Apple Quantity	91.75	–.955	–.200	–.0000618	3.14	30.13	.82	1.12
	(9.61)*	(–3.05)*	(–.40)	(–1.66)				

Note: t-values are in parentheses.
    * indicates significant at the 5% level.

used in an effort to update the model presented in Heydon and O'Rourke (1981). However, the results were not very robust, since reasonable estimates for the coefficients were unobtainable. The nature of the data set had changed markedly since 1977. In particular, the variability of real prices and per capita consumption was very low. Alternative OLS models were attempted, with a variety of substitutes consistent with earlier work, but only pears showed reasonable results as a substitute.

CONCLUSION

Over the past 15 years, Japanese consumption of fresh fruit has declined steadily. Apple consumption has not fallen off as much as some other fruits, but is well below historical highs. Over the next eight years, we project a fairly stable apple market, with possible further small declines in per capita consumption.

Japanese-produced Red and Golden Delicious apples are declining in popularity in favor of such varieties as Fuji, Tsugaru, and Mutsu, which in general provide greater returns to Japanese growers. Japanese retail prices of apples in real terms have been very stable for 15 years. Washington-produced apples are price-competitive with more popular Japanese varieties, but have little price advantage over Japanese-produced Red and Golden Delicious.

More detailed market information is needed to choose appropriate market niches and find ways to make Washington apples competitive in the Japanese apple market. Above all, exporting efforts will have to be carried out with particular concern for quantity and quality, and timing of shipments to maximize the comparative advantages of Washington apples vis-à-vis competition from Japanese apples and other potential imported apples. It is hoped that this report will serve as a basis for industry discussion of future export marketing strategies.

Chapter 13

The Prominence of Japan in the Alaskan King Crab Fishery

Scott C. Matulich
Joshua A. Greenberg
Ron C. Mittelhammer

The Alaskan king crab fishery has undergone dramatic changes in the past two decades, precipitated by unprecedented fishery growth, followed by collapse. Statewide harvests tripled during the 1970s, culminating in a record catch of 180 million pounds in 1980. The dominant fishing area, Bristol Bay, which accounted for 130 million pounds of the record catch, was closed to crabbing only three years later; statewide harvest plummeted to 26 million pounds. Catch continued to decline to 15-17 million pounds in 1988 and 1989.

Throughout these tumultuous two decades, Japan's role in the market for Alaskan king crab has changed from that of no importance to being a primary determinant of king crab prices. Yet, fishery management decisions make no consideration of Japan's implicit or potential role in Alaskan king crab fishery policy.

This paper initially identifies the factors contributing to Japan's prominence in the Alaskan king crab market and then presents an econometric model of wholesale price formation in Japan and the United States. The model deviates substantially from the only other market analysis of the Alaskan king crab fishery. Hanson (1987)– also reported in Matulich, Hanson, and Mittelhammer (1988)–focused exclusively on the domestic market, treated exports as exogenous, and included 1983 (the year of fishery closure) as the terminal year of analysis.

BACKGROUND

Japan's importance to the Alaskan king crab market began in 1972, with the signing of a bilateral treaty that restricted Japanese harvest in U.S. waters. The treaty specified declining annual quotas, ultimately culminating in Japan being barred from the U.S. fishery in 1975. The Magnuson Fisheries Management Conservation Act of 1975 further limited Japanese harvest by extending U.S. jurisdiction to the 200-mile limit. Japan's market share of U.S. harvested product was less than 1 percent in 1974 and 1975. By 1978, Japan. accounted for one-third of U.S. king crab sales. Since that time, its market share has fluctuated between 10 and 50 percent, stabilizing in the late 1980s at around 44 percent.

In the 1970s, the dominant product form sold in the U.S. market switched from canned meats to brine-frozen (cooked) sections. Widespread adoption of brine-freezing technology was motivated by a variety of supply and demand considerations. For example, rising labor costs in Alaska made meat extraction prohibitively expensive; the in-shell, brine-frozen product form (i.e., sections) required one-third the labor. Adoption of brine-freezing technology also was coincidental with rapidly expanding fleet capacity, shorter seasons, and escalating harvests. More crab had to be processed faster, and sections required less processing time than meats. Also, shipping tonnage from Alaska to Seattle increased, allowing for shipment of the bulkier section product. Finally, the brine-frozen product preserved the taste and texture of king crab, and this in-shell product form was preferred by U.S. consumers, particularly those in the "white tablecloth" trade. By 1977, brine freezing was fully adopted and few firms processed meats. Meats now represent only 1 percent of total live-weight product.

Frozen sections have always been the dominant product form sold to Japan. Unlike the U.S. product form, the Japanese prefer raw (uncooked) frozen sections. Secondary reprocessing of the raw frozen crab occurs in Japan. This allows the Japanese firms to reprocess the sections into numerous final product forms that match specific Japanese consumer demand niches. It also allows the Japanese to control quality of the final products. Market dominance of sections makes this product form the only one modeled in this analysis.

Several other market characteristics provide insight into wholesale price formation. The Japanese wholesale market model refers to an ungraded product, purchased FOB Alaska, and shipped in 90-pound bulk containers. The U.S. market, in contrast, refers to a product that is reprocessed in Seattle, size-graded, and packed in 20-pound boxes. Direct shipment to Japan avoids costly transshipment and warehousing in Seattle. But more importantly, it means that Japan enters the market early in the season, requiring processors to commit a portion of their processing facilities to meet Japanese product specifications. Furthermore, the Japanese retail market for king crab appears to be targeted primarily for the end-of-year holiday season, with most of the Japanese purchases occurring in the fourth quarter.

Processors willingly commit processing facilities and live-weight product to the Japanese market during the harvest season for a variety of reasons. First, early-season designation of product to Japan allows processors to avoid shrinkage associated with holding the more perishable raw-frozen processed product in inventory. Second, it allows processors to ship directly from Alaska to Japan, avoiding transshipment costs to Seattle cold storage facilities and also avoiding the capital costs of holding this high-valued product in inventory. Third, early-season sales to Japanese firms increase U.S. prices by lowering available supplies throughout the rest of the year. Fourth, and most importantly, the Japanese pay cash at the time of sale. This provides processors with an early-season cash flow that can be used to offset the cost of unprocessed crab purchases and subsequent reprocessing, holding, and marketing costs. It also lowers processor risk by avoiding possible seasonal wholesale price declines.

The remaining product not shipped directly to Japan is inventoried in Seattle to provide a steady flow of product to the year-long domestic market, with only minor quantities shipped to Japan and all other export markets. All other exports account for only 4.3 percent of total supplies, on average. A small portion of the stored sections is held over into the following season. Some of this crab is held to smooth the transition between harvest/processing in year *t* and year *t*+1 (i.e., to fulfill end-of-season transactional motives). But processors also form a current-period reservation price. If wholesale price falls below the reservation price, the sections are retained in inventory under a speculative motive. Speculative inventory behavior appears to be quite limited partly because of the high opportunity cost of holding such valuable stocks.

PRODUCT ALLOCATION

The above description of the Alaskan king crab market identifies a product allocation scheme that is partially sequential. Japan enters the market early, leaving all other allocations residual to Japan and simultaneously determined. The product allocation and demand structure that follow reflect this characteristic. Rather than arbitraging the alternative markets in order to extract maximum profits across markets and time, processors segment their price formation between the Japanese and U.S. markets. Market segmentation became more formal following the fishing collapse of 1983, which severely disrupted product flow to Japan. Many of the processing firms were purchased wholly or partially by Japanese firms in the mid-1980s. This arrangement further assured Japanese access to product at a "fair" price, where fair refers ostensibly to the margin above exvessel price–the principal cost to the processors.

Equation 1 represents an unusual, simple margin relationship that exists in the Japanese market. Japanese wholesale price (PSECTJ$_t$, dollar/pound) is represented as a margin of average exvessel price (AVEXVP$_t$, dollar/pound).

$$PSECTJ_t = 0.40322 + 1.98705 \; AVEXVP_t$$
$$ (2.57) \qquad (25.16)$$

$$R^2 = 0.980 \; df = 13 \tag{1}$$

Allocation to the Japanese market is implicitly determined by the Japanese demand at the wholesale price determined in Equation 1. Since PSECTJ$_t$ is a function of a predetermined variable, it is segmentable from the rest of the market model. Accordingly, it was estimated using ordinary least squares (OLS) for the period ·1973-87. This simple margin relationship accounts for 98 percent of the variation in Japanese wholesale section price. The parameters are significant at the 0.002 level of significance. Moreover, Figure 13.1 further illustrates the excellent ability of this simple model to track price changes over time.

One should not infer that this specification is a literal representation of market behavior. For example, some brine-frozen product is exported from Seattle to Japan, indicating that limited price arbitraging may occur. However, the influence of these sales and other factors are minor and statistically indiscernible. The specification does suggest, however, that the Japanese may be more concerned with access to product than with price.

Price flexibility coefficients (PFC) provided in Table 13.1 are measures of the sensitivity of wholesale price to changes in the explanatory variable, $AVEXVP_t$. The PFCs at both the 1973-87 mean level and at the more recent 1984-87 mean level indicate that the proportionate change in processors' price offer to Japanese buyers almost matches the proportionate change in $AVEXVPT_t$. At the mean level, a 1 percent increase in exvessel price induces a 0.89 percent increase in processors' price offer to Japan. The slightly higher PFC for the 1984-87 period (0.93 percent) reflects firmer market conditions related to the dramatically curtailed harvests.

The U.S. allocation equation is given in Equation 2 as a quantity-dependent, double logarithmic specification, estimated for the period 1973-87. $SECTUS_t$ (in million pounds) is the allocation of processed sections to the U.S. wholesale market. $PSECTUS_t$ is the New York frozen section wholesale price (dollar/pound), the composite variable ($HARV_t$ − $SECTEXPJ_t$) is statewide harvest ($HARV_t$) net of exports to Japan. The quantity ($SECTHOLD_{t-1}$ + $SECTIMP_t$) is the addition to current supplies from beginning inventories ($SECTHOLD_{t-1}$) and the quantity of Soviet crab imported to the U.S. ($SECTIMP_t$). This alternative supply source is treated as a separate, composite variable because neither it nor $SECTUS_t$ is inventoried beyond the current year. All quantities are measured in millions of pounds.

$lnSECTUS_t = -2.27711 + 1.35942\ lnPSECTUS_t$
(−2.28)   (3.11)

$-0.78218\ lnAVEXVP_t + 0.83286\ ln(HARV_t - SECTEXPJ_t)$
(−2.71)   (8.63)

$+ 0.33826\ ln(SECTHOLD_{t-1} + SECTIMP_t) - 0.42519\ IND83$
(6.17)   (−3.30)

$R^2 = 0.973\ df = 9$ (2)

142

FIGURE 13.1. Actual Versus Predicted Seasonal Average Japanese Wholesale Price F.O.B. for Domestically Produced Frozen King Crab Sections, PSECTJ and PSECTJHAT, Respectively, 1973-1987

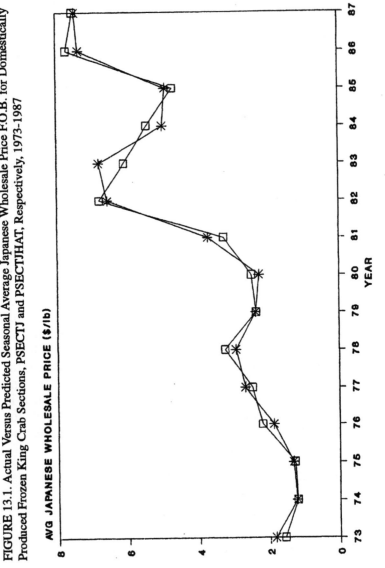

AVG JAPANESE WHOLESALE PRICE ($/lb)

YEAR

□ PSECTJ   ＊ PSECTJHAT

TABLE 13.1. Estimated Japanese Supply Price Flexibility Coefficients

Explanatory Variable	Price Flexibility Coefficients	
Variable	Mean Value (1973 - 1987)[a]	Mean Value (1984 - 1987)[b]
$AVEXVP_t$	0.8921	0.93212

[a] Mean value of $AVEXVP_t$ was used to predict $PSECTJ_t$ and formulate the ratio between $AVEXVP_t$ and $PS\hat{E}CTJ_t$.
[b] The 4 year, 1984-87, average value of $AVEXVP_t$ was used to predict $PSECTJ_t$ and formulate the ratio between $AVEXVP_t$ and $PS\hat{E}CTJ_t$.

Unlike the Japanese market, the allocation of processed sections and the wholesale price are simultaneously determined, requiring the use of two-stage least squares in estimation. The model accurately estimates U.S. wholesale price, as indicated by the reported R^2 and Figure 13.2. All parameters have signs consistent with a priori expectations, and the estimates are significant at the 0.025 level. An indicator variable marked the year of Bristol Bay closure as a structural break in the domestic market allocation behavior.

Constant elasticities for each of the explanatory variables are identical to the coefficients in Equation 2. Supply to the U.S. market is responsive to changes in price, with a direct price elasticity of 1.36.[1] The exvessel price elasticity of −0.78 suggests processors are not highly responsive to increases in exvessel price, reflecting limited market alternatives. A 1 percent increase in processors' principal cost reduces domestic supply by 0.78 percent. Processor responsiveness to changes in exvessel price is most easily compared to that of the Japanese allocation in Equation 1 by deriving the exvessel-price PFC from Equation 2. The exvessel PFC in the domestic market is 0.58. This estimate is substantially below the corresponding Japanese figures (0.89 and 0.93), reflecting processors' consid-

---

1. Conversion of logged predictions to asymptotically unbiased, consistent, and unlogged counterparts requires that the antilog be multiplied by $\exp(0.5 \hat{\sigma}^2)$.

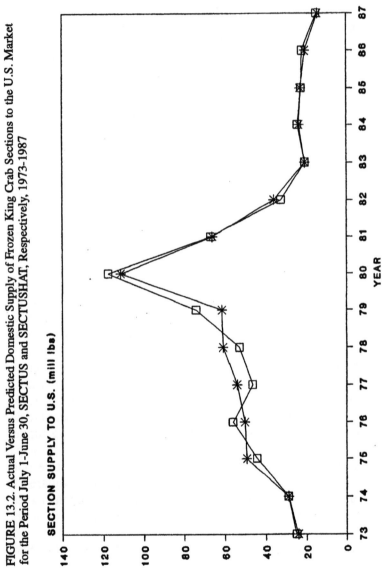

FIGURE 13.2. Actual Versus Predicted Domestic Supply of Frozen King Crab Sections to the U.S. Market for the Period July 1-June 30, SECTUS and SECTUSHAT, Respectively, 1973-1987

144

eration of additional factors (to AVEXVP$_t$) when forming domestic wholesale price offers.

The remaining two elasticities suggest that an increase in the available supply will be allocated primarily to the domestic market. The quantity harvested net of Japan elasticity is 0.83, while the elasticity corresponding to the sum of Soviet imports and beginning inventories is 0.34. One must be careful not to interpret these elasticities as strict allocation proportions, because the actual proportions depend upon the means of the respective variables. Mean SECTUS$_t$ is moderately less than the mean of the current harvest net of Japan, implying that the 0.83 elasticity overstates the allocation proportion. Conversely, the 0.34 elasticity understates the proportion of Soviet imports, as well as the beginning inventories that are allocated to the U.S. market. Mean imports plus beginning inventories are substantially less than mean SECTUS$_t$.

Notice that Equations 1 and 2 together suggest a recursive relationship between the Japanese and U.S. wholesale markets. The Japanese wholesale market has an intraseasonal impact on the U.S. wholesale market, whereas the U.S. market has an interseasonal impact on the Japanese market. An increase in section supply to Japan decreases the net supply available to the U.S. market and, consequently, U.S. wholesale price rises.

Looking solely at Equation 2, the U.S. wholesale market has no apparent effect on the Japanese market. There is, nonetheless, a very important intertemporal feedback between the U.S. and Japanese wholesale markets through exvessel price formation. An increase in the U.S. wholesale price in year t increases the wholesale price expectation of both fishermen and processors in year $t + 1$, causing both to settle on a higher negotiated wholesale exvessel price. Processors then pass along a proportion of this increase to both Japanese and domestic wholesale buyers in $t+1$. Thus, the two markets are interconnected, and any factor that impacts one market will either directly or indirectly impact the other market. A more complete analysis of this type of recursive interrelationship requires an understanding of consumer response to processors' price offer (discussed later in the "Consumer Demand" section).

The final allocation equation models inventory behavior. Processor decisions to hold stocks beyond the current year are motivated ostensibly to meet end-of-year transactional (pipeline) needs. Only limited quantities are held for speculative purposes. Equation 3 was estimated using nonlinear two-stage least squares since the decision to allocate product to inventory is made simultaneously with the decision to allocate product to the U.S. and all other world markets, except Japan.

$$SECTHOLD_t = -1.284459 + 0.119506(HARV_t - MEATPROD_t - SECTEXPJ_t)$$
$$(-0.95) \qquad (6.34)$$

$$+ 0.29548 \ (SECTIMP_t + SECTHOLD_{t-1})$$
$$ (3.52)$$

$$+ EXP(0.46746 \ (PSECTUS_t - PSECTUS_{t-1})) + 12.68999 \ IND77$$
$$ (2.20) \qquad\qquad\qquad\qquad (5.60)$$

$$+ 11.45569 \ IND83$$
$$ (5.00)$$

$$R^2 = 0.934 \ df = 9 \tag{3}$$

where $MEATPROD_t$ is total seasonal king crab meat production. Two indicator variables are included in Equation 3. IND77 removes 1977 as an outlier; U.S. wholesale price increased inexplicably from \$2.68 in 1976 to \$4.01 in 1977, while inventories nearly tripled to record level. IND83 marks the year of the Bristol Bay closure.

This model accurately predicts inventory for the period 1973-87. The specification accounts for 93 percent of the variation in $SECTHOLD_t$, and all parameters–with the exception of the constant term–are significant at the 0.06 level. Figure 13.3 displays the model accuracy.

Total available supplies in Equation 3 are divided into two variables: net supply of sections from current harvest ($HARV_t - MEATPROD_t - SECTEXPJ_t$) and available supplies that are not inventoried for more than the current year ($SECTHOLD_{t-1} + SECTIMP_t$). The model suggests that 12 percent of any change in net supply from current harvest is held in inventory for transactional purposes. Additional supplies that cannot be inventoried ($SECTHOLD_{t-1} + SECTIMP_t$) depress price, motivating processors to inventory additional stock from current harvests.

The third variable of this inventory model represents the limited

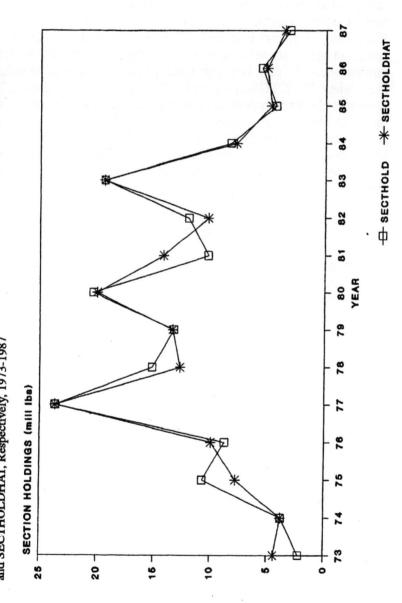

FIGURE 13.3. Actual Versus Predicted End-of-Season Frozen King Crab Section Holdings, SECTHOLD and SECTHOLDHAT, Respectively, 1973-1987

speculative behavior of processors. The difference between current and previous price of sections ($PSECTUS_t$ – $PSECTUS_{t-1}$) captures a naive expectation of future price movements. As long as $PSECTUS_t$ is greater than $PSECTUS_{t-1}$, processors expect an increase in price, stimulating speculative behavior. Processors will diminish stockholding behavior when prices are expected to soften (i.e., when $PSECTUS_{t-1}$ exceeds $PSECTUS_t$). The exponential form was chosen to reflect an increasing (decreasing) rate of speculation as the difference between current and past price increases (decreases). The minor role of speculative behavior is underscored by the price difference elasticities provided in Table 13.2. The elasticities are 0.03 and 0.02, respectively, for the mean and 1984-87 average.

CONSUMER DEMAND

Models of Japanese and U.S. demand for king crab are presented in this section. The king crab prices used in estimation are the same wholesale prices introduced in the "Product Alloca-

TABLE 13.2. Estimated Section Inventory Elasticities

Explanatory Variable	Elasticity	
Variable (X_{it})	Mean Value (1973 - 1987)[a]	Mean Value (1984 - 1987)[b]
($HARV_t$ - $MEATRPOD_T$ - SECTEXPJ)	0.6253	0.3923
($SECTHOLD_{t-1}$ - $SECTIMP_t$)	0.3787	0.6390
($PSECTUS_t$ - $PSECTUS_{t-1}$)	0.0313	0.0245

[a] Mean values for all explanatory variables were used to predict $SECTHOLD_t$ and to formulate the ratio between X_{it} and $SECTHOLD_t$.
[b] The 4-year, 1984-87, average value for all explanatory variables were used to predict $SECTHOLD_t$ and to formulate the ratio between X_{it} and $SECTHOLD_t$.

tion" section (i.e., PSECTJ$_t$ and PSECTUS$_t$), because a retail price series is not available for either the Japanese or U.S. markets. It is assumed that retail and wholesale prices are positively correlated.

Japanese Demand

The previously specified Japanese allocation equation represents Japanese wholesale price as being determined by adding a margin to exvessel price. Consumer response to this price, represented through a demand function, determines the quantity of king crab purchased by Japanese wholesale buyers. The modeling implication of this specification is that wholesale price is exogenous to Japanese consumers. The practical implication is that Japanese wholesalers equilibrate by adjusting quantity, given the processors' price offer. The estimated Japanese demand model is presented in Equation 4 for 1975-87. Estimation began with the 1975 observation because it is the year that Japan ceased crabbing in U.S. waters. Nonlinear ordinary least squares was used in estimation since all explanatory variables are predetermined.

$$\text{SECTCONJ}_t = \text{PSECTJ}_t^{-2.1785} \text{PCEJ}_t^{2.73523} \text{EXCHJ}_t^{-2.37477}$$
$$(-3.87) \qquad (7.25) \qquad (-6.90)$$

$$(\text{IMPR}_t + \text{JHAR}_t)^{-0.58745} \text{SECTCONJ}_{t-1}^{0.27327}$$
$$(-5.00) \qquad\qquad (2.03)$$

$$\exp(-1.02920 \text{ IND8287})$$
$$(-2.43)$$

$$R^2 = 0.991 \; df = 7 \tag{4}$$

Consumption of sections imported from the U.S. (SECTCONJ$_t$), which is set equal to SECTEXPJ in a definitional identity, is estimated as a function of the wholesale price and four demand shifters: Japanese personal consumption expenditures, PCEJ$_t$ (in million dollars); third-quarter exchange rate, EXCHJ$_t$ (yen/dollar); direct substitutes in the form of Japanese harvest of king crab, JHAR$_t$ (in million pounds) plus the quantity of Soviet imports, IMPR$_t$ (in million pounds);[2] and a measure of consumer habit formation, $SECTCONJ_{t-1}$. Price and income are expressed in nominal terms.

The model accounts for 99 percent of the variation in exports to Japan. All parameters conform to a priori sign expectations and are significant at the 0.08 level. The estimation ability of the model is further illustrated by Figure 13.4. The model has some trouble tracking export movements for the 1983-85 period. Record low U.S. harvests for this period may have caused market disruptions that the model did not fully capture.

The exchange-rate variable serves two functions in Equation 4: it converts $PSECTJ_t$ and $PCEJ_t$ to yen values, and it provides a measure of the pure income effect associated with exchange-rate changes. Model performance is enhanced by aggregating U.S.S.R. exports to Japan with Japanese king crab harvest. The significance of the indicator variable (IND8287) suggests that a structural break in demand occurred beginning in 1982. In that year, wholesale price more than doubled from the 1981 level, increasing from $3.28 to $6.80. IND8287 equals one for the 1982-87 period, otherwise it equals zero. The inability to identify statistically a single substitute product probably is related to the fact that Alaskan king crab appears in a wide variety of product forms in the Japanese retail market. Most of these product forms fill narrow demand niches.

The functional form is characterized by constant elasticities, where the parameters are direct elasticities. Several of the elasticities merit comment. As expected, Japanese demand for this luxury good is elastic. A 1 percent change in wholesale price is associated with a 2.02 percent change in Japanese consumption of U.S.-produced king crab. Thus, processors ought not look to raising price as a means of increasing revenue. In the absence of other mitigating market effects, the decline in sales will more than offset gains in price. Demand, however, will expand if, say, the current trend of steadily rising Japanese income continues. Section exports to Japan are estimated to rise (fall) 2.74 percent, for a 1 percent increase (decrease) in personal consumption expenditures.

The elasticity associated with the exchange rate is of special interest because processors appear to target this single factor as the

---

2. Soviet import data is not published. The variable $IMPR_t$ is the quantity imported by the largest Japanese importer.

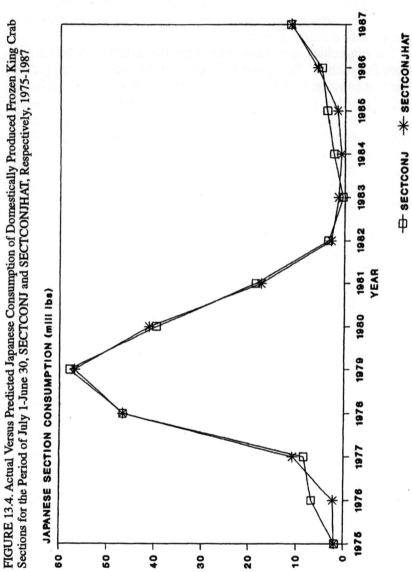

FIGURE 13.4. Actual Versus Predicted Japanese Consumption of Domestically Produced Frozen King Crab Sections for the Period of July 1-June 30, SECTCONJ and SECTCONJHAT, Respectively, 1975-1987

151

most important barometer of Japanese demand. Processors realize a two-faceted exchange-rate effect–a pure income effect and a price effect–whereas Japanese consumers realize only an income effect. As the yen strengthens, the Japanese wholesale price, converted from yen to dollars, increases.

The elasticities associated with these two effects can be derived from Equation 4.[3] Extracting the portion of the exchange-rate parameter that converts $PCEJ_t$ to yen values (recall $PCEJ_t$ was specified in dollars) yields the cumulative price and income elasticity facing processors. The resulting elasticity of –5.110 signals why processors are, and should be, very sensitive to exchange rates. The yen/dollar exchange rate is the most volatile explanatory variable in the demand function, fluctuating more than 35 percent in a single year (1985-86). This volatility translates into substantial movements in the price (dollars) Japanese are willing to pay for king crab.

Isolating the pure income effect of an exchange-rate change is essential to understanding Japanese consumer response to exchange-rate movements. This requires expressing $PSECTJ_t$ in yen. The pure income effect of an exchange-rate change has an elasticity of –3.092. This coefficient means that even in the absence of a decrease in the yen price, a 1 percent decrease in the exchange rate increases Japanese consumption 3.1 percent. Demand shifts outward because of consumer wealth gains.

The Japanese harvest and Soviet import elasticity (–0.587), when considered in conjunction with the low supply level of these products relative to U.S. exports (over most of the historical time period), indicates that these alternative sources of king crab are strong substitutes for domestically produced king crab. Soviet shipments have increased nearly sevenfold between 1979 and 1986. Continuation of this trend may have serious consequences for the U.S. industry. This is especially true if the Soviets improve their processing techniques such that Soviet crab becomes a near-perfect substitute for U.S. crab. The increasing number of U.S.-Soviet joint ventures, coupled with the growing Soviet need for hard currency, suggests

---

3. A P-test represents a Chi-Square test measuring the validity of various non-nested alternative models. For example, the probability of Type I error in rejecting the constant elasticity and linear U.S. models are 0.0002 and 0.993, respectively. The test can result in the acceptance or rejection of one or both models.

that Soviet crab is likely to compete for traditional U.S. markets in Japan. If this occurs, the Alaskan king crab industry is likely to undergo a major structural change.

U. S. Primary Demand

Estimation of a demand function for the U.S. retail market (1970-87) completes the model of consumer demand for Alaskan king crab sections. The demand function models the relationship between U.S. section price, $PSECTUS_t$ (dollar/pound); U.S. section consumption, $SECTCONUS_t$ (in million pounds); and demand shifters. $SECTCONUS_t$ is set equal to domestic supply ($SECTUS_t$) in a definitional identity.

The inverse demand function is given in Equation 5. All prices and income are specified in nominal terms. Simultaneity between $PSECTUS_t$ and $SECTCONUS_t$ requires the use of two-stage least squares in model estimation.

$$PSECTUS_t = -2.06135 - 0.04361 \, SECTCONUS_t + 0.00152 \, PCE_t$$
$$\quad\quad (-2.41) \quad\quad (-4.46) \quad\quad\quad\quad\quad (1.72)$$

$$+ \, 2.15540 \, PLOB_t + 0.01501 \, SECTCONUS_{t-1}$$
$$\quad (2.35) \quad\quad\quad (1.73)$$

$$+ \, 1.25166_t \, IND83$$
$$\quad (1.84)$$

$$R^2 = 0.972 \; df = 12 \quad\quad\quad\quad\quad\quad\quad\quad\quad\quad (5)$$

PCE_t (in million dollars) is U.S. personal consumption expenditures. $PLOB_t$ is an exvessel American lobster price index. $SECTCONUS_{t-1}$ proxies for habit formation.

All parameters conformed to a priori sign expectations. The ability of the model to estimate wholesale price changes is reflected by the reported R^2 and Figure 13.5. The model had the same difficulty in tracking price changes in the early years, when brine-frozen sections were being introduced to the American market.

The U.S. and Japanese demand models differ in functional form. The validity of the different specifications was tested with a P-test (Davidson and MacKinnon, 1981). Specifically, the linear functional

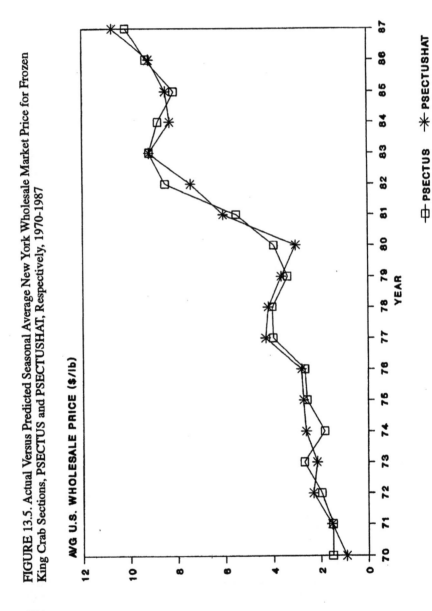

FIGURE 13.5. Actual Versus Predicted Seasonal Average New York Wholesale Market Price for Frozen King Crab Sections, PSECTUS and PSECTUSHAT, Respectively, 1970-1987

154

form was tested versus its constant elasticity counterpart for each market. The multiplicative form was rejected in the U.S. market while the linear form was accepted. Conversely, the linear form was rejected in the Japanese market, while the multiplicative form was accepted.

Price flexibility coefficients for each of the explanatory variables are provided in Table 13.3. As before, two coefficients are presented for each variable: one calculated at the mean of the explanatory variable and one calculated at the average value for the period 1984-87.

The section consumption PFCs reveal price inflexibility at both the mean, PFC = −0.359, and the 1984-87 average value, PFC = −0.101. The very low PFCs suggest that even small price increases will dramatically depress domestic sales. King crab has become so expensive that consumers are highly responsive to price changes. Processors have little latitude in adjusting price. This is, in part, due to the substitutability of more abundant and less expensive lobster. American lobster price PFCs are 1.164 and 0.831. The lower PFC for the 1984-87 period probably reflects the firmer lobster market during the latter period.

The personal consumption expenditure PFCs are relatively constant when evaluated at the two mean points on the inverse

TABLE 13.3. Estimated U.S. Demand Price Flexibility Coefficients

Associated Explanatory	Price Flexibility Coefficients	
Variable (X_{it})	Mean Value (1970 - 1987)[a]	Mean Value (1984 - 1987)[b]
SECTCONUS$_t$	-0.3587	-0.1007
PCE$_t$	0.4567	0.5093
SECTCONUS$_{t-1}$	0.1191	0.0374

[a] Mean values for all explanatory variables were used to predict PSECTUS$_t$ and formulate the ratio between X_{it} and PSÊCT$_t$.
[b] The 4-year, 1984-87, average values for all explanatory variables were used to predict PSECTUS$_t$ and formulate the ratio between X_{it} and PSÊCTUS$_t$.

demand function (0.509 and 0.457, respectively). These price flexibilities indicate that changes in income have a significant impact on domestic king crab price.

DEFINITIONAL IDENTITIES

Two definitional identities–shown in Equations 6 and 7–complete the wholesale price formation model specification by setting king crab consumption in the U.S. and Japan ($SECTCONUS_t$ and $SECTCONJ_t$) equal to the supply of domestically produced king crab allocated to each country ($SECTUS_t$ and $SECTEXPJ_t$).

$$SECTCONUS_t = SECTUS_t \tag{6}$$

$$SECTCONJ_t = SECTEXPJ_t \tag{7}$$

SUMMARY, CONCLUSIONS, AND IMPLICATIONS

The supply allocation and demand framework represents a deviation from conventional arbitraging behavior. Japan and the U.S. wholesale markets operate in a recursive fashion, with Japan entering the market first. This early entry provides substantial benefits to processors. First, the pricing structure to Japan assures a "fair" margin above processors' costs. Second, it provides an early-season cash flow. Third, it decreases available supply to the domestic market, thereby firming U.S. price.

Model results indicate that processors properly focus on exchange-rate movements. The high elasticity associated with the exchange rate indicates that Japanese king crab consumption is very sensitive to changes in this variable. In fact, much of Japan's increasing market share can be attributed to the declining exchange rate (yen/dollar). The results also suggest that U.S. processors should be concerned with the steady growth in Soviet crab exports to Japan, since Soviet crab is a close substitute for the U.S. product.

Despite the fact that the U.S. market is the largest outlet for king crab, it is a residual market to Japan. The Japanese enter the market early, allowing for segmentation of the Japanese market from all other outlets. Accordingly, the U.S. wholesale price is responsive to changing market allocations–it is not determined by a margin relationship with exvessel price.

U.S. demand results are an indicator of consumer reaction to future price increases. Very low price flexibility suggests recent high prices may be nearing choke prices, where domestic consumers will stop purchasing king crab. In other words, further price increases may cause precipitous declines in domestic demand. This observation portends a potential structural change in the market if exchange rates fall sufficiently. A falling exchange rate (yen/dollar) should expand Japanese demand and, thus, wholesale price in both markets. The Japanese market could conceivably become the only effective market for Alaskan king crab. The importance of Japan in future king crab management decisions cannot be overemphasized.

Chapter 14

Washington Wagyu: A Beef Breed for Japanese Tastes

Raymond A. Jussaume, Jr.
Raymond W. Wright, Jr.
Jerry J. Reeves
Kristen A. Johnson

INTRODUCTION

On April 1, 1991, the importation of beef into Japan became liberalized. The direct role that the quasi-governmental Livestock Import Promotion Council (LIPC) played in beef imports by purchasing foreign beef and then selling it in the domestic market was to be eliminated. The current program of import volume quotas, which is administered by the LIPC, will be replaced by a tariff system. This means that any Japanese buyer can freely purchase foreign beef, but those purchases will be subject to a 70 percent tariff. All such purchases will be subject to a 60 percent tariff beginning on April 1, 1992, and a 50 percent tariff commencing on April 1, 1993. Predictions abound on what impact this dramatic change will have on how Japanese firms buy imported beef, on the international beef trade, Japanese domestic beef production, and Japanese consumer demand for beef. There can be no doubt, however, that a metamorphosis is forthcoming both in the Japanese beef market and in how United States producers fit within this market. The livelihood of a number of United States businesses will depend on accurately anticipating and preparing for those transformations.

When the beef trade agreement was first announced in 1988,

euphoria spread over the United States beef industry. Bolstered by economic analyses that predicted Japanese beef imports would expand anywhere from four to ten times in the first few years of the new free-trade regime (Ohga, 1988; Sanderson, 1987), U.S. beef-industry representatives began announcing to American producers that "between now and the end of the century, the Japanese beef market will be the largest growth market for beef consumption in the world" (Seng, 1989). As a result, many Americans have come to view the opening of the Japanese market as an easy opportunity for expanded beef sales.

While the opening of the Japanese beef market offers tremendous promise, enough evidence already exists to demonstrate that expanding American beef exports to Japan will not be achieved effortlessly. Although per capita beef consumption in Japan has risen more than fourfold in the past three decades (Table 14.1), it may be imprudent to claim that "The attributes of U.S. high-quality beef are similar to that of Japanese domestic beef" (Middaugh, 1990) and that Japanese consumers will flock to purchase imported U.S. beef in large quantities. Unfortunately, many in the U.S. beef industry have assumed this scenario will come to pass and have failed to develop either a market research program to evaluate how U.S. beef fits into the Japanese marketplace or a strategy for adapting the U.S. product to make it more desirable in the eyes of Japanese consumers.

This report describes the efforts of a team of researchers who have confronted this deficiency by developing a project to blend Wagyu (pronounced "Wah-gyou") genetics and Japanese-style feeding practices into an American system to furnish beef products more suited to Japanese tastes than traditional U.S. beef exports. This work is being sponsored by Washington State University's IMPACT (International Marketing Program for Agricultural Commodities and Trade) Center, whose mission is to promote scientific research that can augment the State of Washington's ability to compete in international agricultural markets.

CONSUMER DEMAND FOR BEEF IN JAPAN

The first, and most crucial, step in an effective marketing strategy is to understand consumer needs and wants and to move toward

TABLE 14.1. Annual Per Capita Japanese Consumption of Livestock Products and Fish

(in Kilograms[1])

Year	Beef	Pork	Chicken	Fish
1960	1.2	1.3	0.4	28.1
1965	1.5	3.0	1.9	29.2
1970	2.1	5.3	3.7	31.6
1975	2.5	7.3	5.3	34.9
1980	3.5	9.6	7.7	34.8
1985	4.4	10.3	9.1	35.8
1988	5.4	11.4	10.5	37.0

[1]Net edible weight.

Source: Ministry of Agriculture, Forestry, and Fisheries (MAFF), Food Balance Sheet, various issues.

promoting consumer satisfaction (Moss and Richardson, 1985; Smith, 1987; Stern and Sturdivant, 1987). It is impossible to be successful in business without giving customers the product they want. Similarly, any explanation of a research project geared towards producing a type of beef more suited to Japanese tastes must begin with an analysis of Japanese consumer demand for beef.

For hundreds of years, the Japanese diet consisted primarily of rice, barley, soybean products, vegetables, and some fish. Only a century ago did the eating of meat from four-legged animals begin to become socially acceptable. Prior to that time, religious prohibition and resource constraints were among the reasons why the average Japanese did not consume any animal flesh. This historical tradition helps explain why it has only been during the post-World War II era that beef and other animal protein consumption have begun to become a notable part of the Japanese diet (Table 14.1).

Since the mid-1960s, meat consumption in Japan has nearly

tripled. However, beef consumption has not risen as fast as that of other meats. This trend may be explained by a number of factors, including the slow expansion of the domestic beef industry, government control of beef imports, and the atypical eating preferences of many Japanese consumers. The latter is exemplified by the fact that Japanese consumers continue to eat much less meat and more fish than consumers in other industrialized countries (Table 14.2). This is quite understandable given Japan's vegetarian heritage, which permitted fish consumption, and its proximity to maritime resources. Over the course of the present century, Japanese consumers have begun to eat and prefer varieties of fish, such as raw tuna, that were considered to be less desirable in centuries past (Williams, 1986). In other words, the modernization of the Japanese diet has been characterized more by an increase in fish consumption than by a surge in beef consumption.

Just as the modernization of the Japanese diet has not been typified by the rise in beef consumption seen in other industrialized countries, so too the pattern of beef consumption in Japan is very different from that observed elsewhere. For example, IMPACT-sponsored research (Jussaume, 1989) has demonstrated that Japanese residents of Kobe consume beef *more frequently* than the residents of Seattle, Washington (Figure 14.1). The explanation for this surprising result is that Japanese cuisine emphasizes a greater variety of food products, with smaller portions, compared with American cooking. Small portions of beef and other animal proteins are frequently added to soups, fried vegetables, and other dishes. Meat is rarely served in large portions or in a simple entree accompanied by a serving of a single vegetable and potatoes, as might be done in many American households.

Perhaps the most preferred style of cooking beef in Japan is in the dishes shabu-shabu and sukiyaki. In both cases, the meat is sliced very thinly and is placed in boiling water along with a variety of vegetables. These two food preparations resemble, and were probably patterned after, traditional methods of cooking vegetables and fish. Shabu-shabu and sukiyaki require the use of heavily marbled beef to maintain tenderness during the boiling process. Beef recipes such as these, as well as the fact that small amounts of beef are used almost as a garnish in many dishes, illustrate why

TABLE 14.2. Consumption of Meats and Fish of Selected Countries, 1985

(in Kilograms[1])

Country	Annual Per Capita Consumption Meat	Fish
Canada	96.5	7.2
West Germany	99.8	6.4
Japan	25.1	35.8
United States	117.5	7.1

Source: OECD, Food Consumption Statistics

heavily marbled, tender, delicate-looking cuts of beef are preferred by Japanese consumers.

Americans are habitually informed by the media that beef in Japan is many times more expensive than beef in their own country. As Lin and Mori (1990, p. 4) have pointed out, however, this is too simplistic an appraisal:

One can easily find beef priced at $40 to $50 per pound in ordinary supermarkets and even up to $100 per pound in many department stores. However, this comparison [with American beef] is unfair due to quality differences between various types of beef in Japan. Second grade dairy beef, the most popular [Japanese] domestic beef which *may* [emphasis added] correspond to U.S. Prime or high Choice grade, is priced around $4 per pound, carcass weight, at wholesale markets.

This price difference is easily observable at the Japanese retail level, with imported beef priced approximately $8 to $10 per pound less in supermarkets and butcher shops than virtually all cuts of domestic dairy beef. In spite of this price advantage, imported beef was considered to be of an inferior grade by many Japanese consumers.

Simply put, the quality characteristics of beef that appeal to Japanese consumers are not identical with those preferred by most Americans. Japanese shoppers express this desire with a willing-

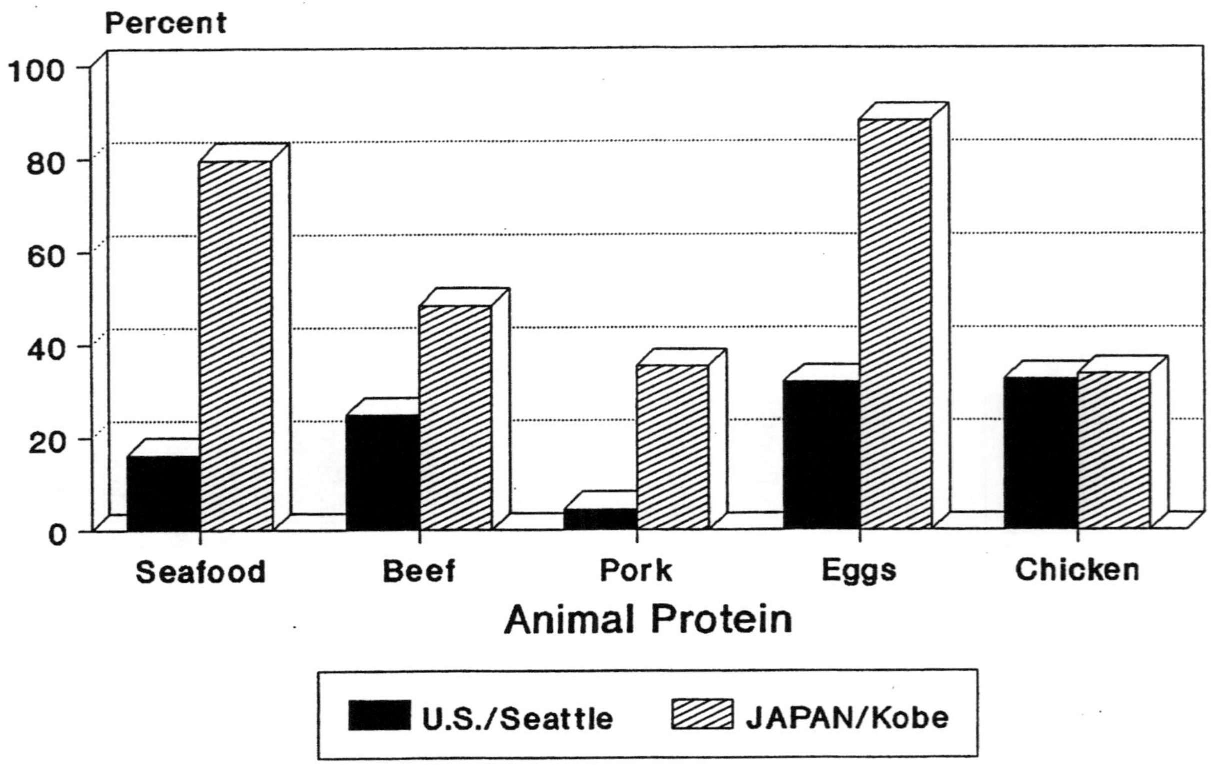

FIGURE 14.1. Percentage of Households Where Protein Is Eaten Three Times a Week or More

ness to pay higher prices for most cuts of domestically produced beef. The overseas producer who can provide a product having some of the quality attributes preferred by Japanese consumers will be more successful in exporting his product to Japan.

WHY WASHINGTON STATE WAGYU?

There are two types of beef produced in Japan. The first, known as Wagyu, is a type of cattle distinct to Japan. The four modern Japanese breeds known collectively as Wagyu (Japanese Black, Japanese Brown, Japanese Poll, and Japanese Shorthorn) are the historical result of having cross-bred indigenous Japanese cattle, which were traditionally used as a draft animal, with a variety of European breeds during the early part of the twentieth century (Longworth, 1983). The meat from this animal is valued for its taste and its extreme tenderness, due in large part to the Wagyu's ability to marble extensively with little backfat. It is the opinion of the Washington State University (WSU) Wagyu research team that the Japanese Wagyu cattle *do* marble much better than U.S. cattle and that the U.S. cattle industry does not have at its disposal the genetic technology to produce a carcass that marbles to the level of the Japanese Wagyu–even with the use of a Japanese feeding system.

The only other kind of beef produced in Japan is fed dairy beef. Second-grade dairy beef ostensibly corresponds to U.S. Prime or high Choice grade. However, there is reason to doubt that they are equivalent, since most imported beef is not officially evaluated by the Japan Meat Grading Association. Extensive research by Dr. Hiroshi Mori and his colleagues has clearly demonstrated that the types of beef imported into Japan up to the present are quite different from domestic beef, including dairy beef (Mori, Lin, and Gorman, 1989a, 1989b). This difference is undoubtedly due in large part to the fact that Japanese beef farmers, including those that fatten dairy beef, are producing for a beef market unlike that in the United States.

The distinctiveness of the Japanese beef market can be illustrated through a brief description of its grading system.[1] The Japanese have three yield grades with A being the highest, followed by B and C.

---

1. The following description is adapted from Lin and Mori, 1990.

The quality grades range from a high of 5 to a low of 1. Thus, there are 15 different Japanese meat grades. The quality grade is based on: (1) marbling, (2) meat color and brightness, (3) meat firmness and texture, and (4) fat color, luster, and quality. The minimum grade of any one of these four characteristics is the overall quality grade. For example, a carcass can grade 5 in marbling, color, and firmness, but if it has a 3 in fat color, then the overall carcass quality grade will be 3. This grading system is relatively new, having been instituted in 1988 to replace an older, more subjective predecessor. The reason commonly given for the introduction of the new system was that it would discourage the long feeding periods of beef animals in Japan. Yet the Japanese market continues to discriminate in favor of beef carcasses that have a high degree of marbling. This is a further indication that the Japanese consumer discriminates in favor of highly marbled cuts in purchasing beef.

Japanese consumers are willing to pay higher prices for beef that meets their own particular quality standards. Degree of marbling is one of the most important determinants of this quality standard. However, while marbling can be increased with longer feeding periods, altering feeding practices is insufficient in and of itself to ensure a high-quality carcass for the Japanese market. Many individual animals will not marble well, despite extended feeding periods, and breeds vary in the percentage of animals that express this trait. In addition, it is crucial in the Japanese marketplace to achieve extensive marbling with a minimum amount of backfat, a characteristic generally not found in U.S. cattle breeds. This attribute is found in some, but not all lines, of the Japanese Wagyu cattle.

As mentioned previously, there are four commercial breeds of Japanese Wagyu cattle.[2] The Japanese Shorthorn, partly descended from European breeds, is solid red in color and is most popular in the northern part of Honshu island. The Japanese Poll breed, which has some Angus ancestry, is really just a curiosity, with only about 2,000 females existing in Yamaguchi Prefecture.[3] The Japanese Red, which is apparently related in part to the Korean Kangyu breed, is found mainly in Kumamoto and Kochi Prefectures. The

---

2. This paragraph is adapted from Longworth, 1983.
3. A prefecture is analogous to a state in the United States.

Japanese Black, the most popular of the Wagyu cattle breeds, is found in all areas of Japan and makes up about 90 percent of beef cattle in Japan. The Japanese Blacks consist of three main strains: the Tottori, Tajima, and Hiroshima. The Japanese Blacks, particularly the Tajima strain, are known for their ability to marble. It is this characteristic more than any other that accounts for the high demand and price paid for Wagyu carcasses in Japan.

In 1976, four Wagyu bulls, two black and two red, were imported into the U.S. As far as can be determined, these are the only Wagyu animals ever to be exported out of Japan. These four bulls are all dead, but there is still semen available from them. A number of cattle with 3/4 to 31/32 Wagyu blood are owned by several producers in the United States. Given the unique aspects of the Japanese market for beef and the genetic capability of the Wagyu breed to produce a type of beef favored by many Japanese consumers, it was the opinion of the WSU research team that an effort needed to be made to identify the marbling capability of the Wagyu genetics available in the United States and to incorporate that trait into compatible U.S. cattle breeds. To this end, WSU purchased one 15/16 Wagyu bull, two Wagyu heifers, and four long yearling Wagyu steers in 1989, and additional 17 percentage Wagyu animals in 1990.

CURRENT RESEARCH STATUS

As a part of the research program designed to evaluate the use of Wagyu or percentage-Wagyu cattle in Washington beef production systems, the four Wagyu steers were matched with four Angus and four Longhorn steers of approximately the same age, and are now being fed identical feed rations in a system approximating Japanese feeding practices. Cattle are fed at a level designed to produce approximately 1.5-2 pounds of gain per day by the Wagyu cattle. The diet is limited grain (rolled barley, dehydrated alfalfa, corn, trace mineralized salt, and vitamins) and ammoniated wheat straw. Six months prior to the attainment of slaughter weight, the grain portion of the diet will be greatly increased and all carotenoid containing feeds will be dropped from the diet to keep the fat white. Daily intakes, monthly body-weight gain, body composition (urea space), and ultrasound profiles will be measured over the course of the experiment. Fat and muscle biopsies will also be taken.

One of the main questions to be answered by this part of the Wagyu study is how these Japanese cattle grow and develop. Information in English on the performance and requirements of the Japanese Black breed is limited. Available information indicates that the energy requirements for maintenance are lower for the Wagyu than for other breeds of cattle. Confirmation of this hypothesis is crucial for determining the desirability of incorporating Wagyu genetics into U.S. cattle breeds. Other questions will also be addressed by this line of research: Are patterns of growth seen with Wagyu different than those of British or Continental breeds? How will the Angus and Longhorns marble, in comparison with the Wagyu, when grown and finished on a slow feeding regime? Preliminary results (Figure 14.2) indicate that when fed identical diets, the Wagyu steers have an average daily gain and feed-efficiency rate that is superior to the Longhorn but inferior to the Angus.

Another line of research in the WSU Wagyu project will determine which U.S. breed crosses best with Wagyu. In utilizing the genetics of the Wagyu cattle for export, we must cross them with our own breeds of cattle. It is thus vital to evaluate the carcass characteristics of the progeny of crosses made by using the major available beef breeds and the Wagyu. This is an impossible task to perform with the limited number of cattle available at WSU. Therefore, the Wagyu research team is working with Cooperative Extension agents to distribute Wagyu semen from WSU's bull to Washington State producers; the semen will be used to breed with six different types of U.S. cattle. WSU will purchase back one-half of these first generation calves, along with an equal number of contemporary straight-bred calves. In addition, embryo transfers with the two Wagyu heifers are being performed in order to produce more seed stock and animals for use in the carcass evaluation line of research. Over ten pregnancies had been achieved by Fall 1990.

· In order for producers to make the most efficient use of the Wagyu genetics currently available in the United States, it will be helpful to isolate those Wagyu bulls in U.S. that have offspring with the highest marbling ability. The four original Wagyu bulls shipped into the U.S. are dead, and their semen supply is limited. We assume

FIGURE 14.2. Wagyu Project–Cattle Performance

Breed	ADG (lb/d)	FE (lb feed/lb gain)
Wagyu	1.45	13.92
Angus	1.68	11.40
Longhorn	1.30	14.30

the U.S. will not be able to import any additional Wagyu semen from Japan in the foreseeable future. Thus, the descendants of the four original bulls must be evaluated for marbling ability. If genetic differences exist between bulls, this would have a bearing on which bulls were selected as herd sires. Knowledge of genetic differences will also determine how crossbred females would be selected in the propagation of high-marbling lines of Wagyu cross cows. WSU has been granted access to semen from 21 percentage Wagyu bulls. A number of cattlemen have expressed a willingness to have their cows bred to these bulls and retain ownership of the offspring during the feeding period, after which the WSU research team will collect carcass data from the progeny and evaluate which are the superior carcass bulls.

Another element of the WSU Wagyu research project will be to ascertain the break-even price on a crossbred carcass that has been fed for extended periods. The goal of the production research discussed above is to identify the achievable grades of carcasses for different U.S. breeds, Wagyu crosses, and Wagyu cattle fed different rations and at varying times. By integrating marketing and production research, the most economically efficient combination of breed crosses and feeding durations can be determined. Both the feed-conversion ratios of Wagyu cattle and the data on Angus cattle that are being fed for the Japanese market in the state of Washington will be evaluated. These data will be used to estimate the cost of longer feeding, which will be compared with results of break-even analyses to derive the profitability estimates of longer feeding using various percentage Wagyu cross.

Finally, research is ongoing with colleagues at WSU's sister University in Japan, Nihon University, to continuously monitor changes in the Japanese market for beef products. The only accurate prediction that can be guaranteed at this time regarding the opening of the Japanese beef import market in April 1991 is that monumental changes will be taking place and that competition in the Japanese marketplace will be fierce. For the WSU Wagyu research project to succeed, it is imperative that consumer preferences, market structure, and prices be constantly monitored. This information will be used by production-oriented researchers to ensure that the type of cross-bred cattle produced by the WSU team is one that can be efficiently produced in Washington State and sold for an adequate price in Japan.

DISCUSSION

While predicting the future is always risky, and foreseeing the future of the Japanese beef market is even more so, the following first-hand reports on recent developments in the Japanese beef market are insightful in understanding the force that may come into play.[4]

> Stocks of imported beef held by meat processors are rising as a result of sluggish growth in retail demand. According to industry estimates, private stocks of beef have broken through the 60,000-ton level to reach a record high. The growth of beef sales at supermarkets and butcher shops that handle imported beef has slowed since April [of 1989]. Industry figures indicate that the consumption of uncooked beef has grown only 3% in the first six months of 1989 compared with the previous corresponding period, and has fallen below the growth rate of 1988.
>
> The retail price of some imported sirloin has also fallen, reflecting a gradual rise in supply. According to a survey by meat processors, the shipment price for imported, high-quality

---

4. The reader should bear in mind that then existing import quotas represented a *requirement* on the part of the LIPC to purchase the entire volume of beef allowable under the U.S.-Japan beef agreement.

cuts of beef sirloin in September was about 1,950 [yen] per kilogram ($6.17 per pound), a 12% drop on the February price of 2,200 [yen] per kilogram ($6.96 per pound). (*Japan Agrinfo Newsletter*, 1989)

"Gone is the power of the words 'imported beef' to attract customers," is what some supermarket buyers responsible for cut meat are saying. Despite its price being somewhat higher, *Wagyu* is selling well at the dressed meat counters of supermarkets, where neatly sliced meat in packs is set out in display.

In 1988, import liberalization of beef was decided upon, and consumer interest focused on imported beef. Up until the [1989] Christmas season, department stores and supermarkets built their sales promotions around imported beef. However, as beef consumption increased, "high-priced but tasty beef has come to be preferred by customers." The year-end emphasized this trend even more strongly, with increased sales of materials for high-grade cooking, such as steak, *shabu-shabu*, *sukiyaki* and so on. From the middle of December last year [1989], the number of shops concentrating their sales efforts on *Wagyu* has been increasing. (*Japan Agrinfo Newsletter*, 1990)

These observations lead to an inescapable conclusion. While Japan does hold promise as a growing market for the existing U.S. beef and cattle industry, this opportunity is not without risks. While the WSU Wagyu research project cannot guarantee increased sales of Washington-produced beef to Japan, it does represent a conscientious attempt to do what needs to be done. That is, to recognize the unique aspects of the Japanese beef market and to make attempts to produce the type of beef that can be competitive in that particular marketplace.

Chapter 15

Edamame:
The Vegetable Soybean

John Konovsky
Thomas A. Lumpkin
Dean McClary

INTRODUCTION

Edamame is a specialty soybean (*Glycine max* L. Merrill) harvested as a vegetable when the seeds are immature (R6 stage) and have expanded to fill 80 to 90 percent of the pod width. Like field-dried soybeans, the seeds of edamame varieties are rich in protein and highly nutritious. Worldwide, it is a minor crop, but it is quite popular in East Asia.

Edamame is consumed mainly as a snack, but also as a vegetable, an addition to soups, or processed into sweets. As a snack, the pods are lightly cooked in salted, boiling water and then the seeds are pushed directly from the pods into the mouth with the fingers. As a vegetable, the beans are mixed into salads, stir-fried, or combined with mixed vegetables. In soup (*gojiru* in Japanese), the beans are ground into a paste with miso, which is used to form a thick broth. Confectionery edamame products, such as sticky rice topped with sweetened edamame paste (*zunda mochi* in Japanese), are occasionally prepared. For marketing, edamame pods are sold fresh on the

Special thanks for their help go to Dr. N. Kaizuma and Dr. Y. Takahata at Iwate University; Mr. Y. Kiuchi at the Iwate Prefectural Agricultural Experiment Station; Dr. H. E. Kauffman, Dr. R. Bernard, and Dr. D. Erikson at INTSOY; and Dr. S. Shanmugasundaram at AVRDC.

stem with leaves and roots, or stripped from the stem and packaged fresh or frozen, as either pods or beans.

HISTORY

China. Edamame (*mao dou*) use was first recorded around 200 B.C. as a medicinal (Shurtleff and Aoyagi, unpublished) and is still very popular (Jian, 1984). Major varieties (and their production areas) include: *Sanyuewang* in Zhejiang Province; *Wuyuewu* around Shanghai and Nanking; *Wuyueba* near Hangzhou; *Baishuiou* in the Chengdu area; *Liuyueba* around Hefei, Wuxi, and Hangzhou; *Baimaoliuyuewang* and *Daqingdou* near Nanking; and *Jiangyoudou* around Shanghai (Guan, 1977). Numerous land races are still cultivated, particularly around Shanghai and in Jiangsu.

Japan. Though soybeans were introduced from China at an early date, the first recorded use of edamame is the description of *aomame* in the *Engishiki* (927 A.D.), a guide to trade in agricultural commodities. It describes the offering of fresh, podded soybean stems at Buddhist temples (Igata, 1977). Early interest in the edamame crop was seasonal, and it climaxed with the viewing of the full moon in September and October (Shurtleff and Aoyagi, unpublished).

Historically, edamame was grown on the bunds between rice paddies, but with the current rice surplus and official pressure to convert paddy fields to other uses, field production is more common (Gotoh, 1984).

Japan is the largest commercial producer of edamame–turning out nearly 105,000 tons in 1988 (MAFF, 1990)–and it is also the largest importer of the bean, bringing in over 33,000 tons in 1989 (JTA, 1989). Taiwan supplies over 99 percent of those imports as frozen edamame. Almost all Japanese production is consumed as fresh product during the summer months (Kono, 1986).

Korea. Morse noted edamame's *(poot kong)* availability in 1931 (Shurtleff and Aoyagi, unpublished). It is still cultivated throughout the country and variety development is underway (Hong, Kim, and Kwang, 1984), with research on crop-management options (Lee, 1986a, 1986b).

North America. Edamame is known by many names in North

America (Shurtleff and Aoyagi, unpublished). The most common is vegetable soybeans, but also beer beans, edible soybeans, fresh green soybeans, garden soybeans, green soybeans, green-mature soybeans, green vegetable soybeans, immature soybeans, large-seeded soybeans, vegetable-type soybeans, and the Japanese name, edamame. Reference to the color green is confusing because mature soybean seeds with a green seed coat (or cotyledon) are also called green soybeans.

Other Countries. Other countries that have produced commercial quantities of edamame include Argentina, Australia, Israel, Mongolia, New Zealand, and Thailand. Home gardeners are known to produce it in Bhutan, Brazil, Britain, Chile, France, Germany, Indonesia, Malaysia, Nepal, Philippines, Singapore, and Sri Lanka (Wang et al., 1979).

Edamame Research. Research has been conducted for over 50 years. Dorsett and Morse collected extensive germ plasm during 1929-31, and Morse used it to develop 49 varieties of edamame (Hymowitz, 1984). Research flourished during the 1930s and 1940s because of a protein shortage (Smith and Van Duyne, 1951). The University of Illinois tested palatability (Woodruff and Klass, 1938) and regional adaptation (Lloyd, 1940) of many soybean varieties, some of which were true edamame, and many companies experimented with canning fresh beans (Shurtleff and Aoyagi, unpublished).

A second surge of interest began with the rise of organic agriculture in the 1970s. The Rodale Research Center focused on adaptability and quality (Hass, Gilbert, and Edwards, 1982). Basic agronomic research was begun at Cornell (Kline, 1980) and seed companies developed new varieties (e.g., butter beans). Today, some home gardeners grow edamame, but there is little commercial production. Asian-Americans seeking edamame are usually limited to frozen imports in specialty supermarkets.

QUALITY

At harvest, edamame has lower trypsin-inhibitor levels, fewer indigestible oligosaccharides, and more vitamins than field-dried soybeans (Rackis, 1978). True edamame varieties are not easy to distinguish from immature, grain soybeans, except for a few unique characteristics (Rackis 1981). The larger seeds of edamame are considered superior to soybeans in flavor, texture, and ease of cooking, but significant differences in chemical composition have not been identified. Phytic-acid levels are higher (Gupta, Kapoor, and Deadhar, 1976) and may help explain why edamame is more tender and cooks faster. Large-seeded soybeans have nutritionally superior protein (Hymowitz, Mies, and Klebek, 1971), e.g., the gain-per-feed ratio was higher for edamame fed to rats than for soybeans (Yen et al., 1970). Edamame has a slightly sweet, mild flavor and nutty texture, with less objectionable beany taste. It has larger, easily shattered pods, seeds with a fragile seed coat (Smith and Van Duyne, 1951), and stems that may have several podless nodes (Shanmugasundaram, Tsou, and Cheng, 1989).

Edamame protein levels were thought to be slightly higher than soybeans (Liener, 1978; Smith and Circle, 1978), but recent research indicates larger seeds have a higher percentage of oil and less protein than smaller seeds (Reddy et al., 1989). Since protein and total sugars are negatively associated (Hymowitz et al., 1972), too much protein leads to a lack of sweetness, an important component of flavor. Conversely, oil content and total sugars are positively associated, but an excessively oily taste is unacceptable.

Edamame quality is evaluated by distributors and consumers for appearance, aroma, flavor, and firm texture after cooking. Morphologically, edamame pods should have white pubescence (Watanabe, 1988), preferably sparse and soft (Sunada, 1986); the hilum should be light brown or gray; the pods must have two or three seeds; most pods should be at least 5 centimeters long; 500 grams should contain less than 175 pods; 100 seed-weight should exceed 30 grams (Shanmugasundaram, Tsou, and Cheng, 1989); the pods should be completely green, with no hint of yellowing (IDA, 1990); the mature seed coat can range from yellow or green to brown or black (Kiuchi et al., 1987); and the pods must be unblemished.

In the Iwate Prefecture (IDA, 1990), grade A edamame must have 90 percent or more pods with two or three seeds. The pods must be perfectly shaped, completely green, and show no injury or spotting. Grade B edamame must have 90 percent or more pods with two or three seeds, but can be lighter green, and a few pods can be slightly spotted, injured, malformed, short, or have small seeds. In both grades, pods cannot be overly mature, diseased, insect-damaged, one-seeded, malformed, yellowed, split, spotted, or unripe. Pod color is the most visible quality of edamame. In addition to varietal influences, several management factors affect color. In general, exposure to sunlight has the greatest positive influence, followed by moisture level and supplemental fertilizer. When leaves were removed at two-thirds of mature height, the pods darkened considerably over the control by harvest due to increased exposure to sunlight (Chiba, Yaegashi, and Sato, 1989); however, researchers at the Asian Vegetable Research and Development Center (AVRDC) have suggested lower leaf area increases sunburn (Shin, 1988). The pods seem to darken with the increased chlorophyll content caused by lower moisture levels (AVRDC, 1988). Post-anthesis application of phosphorus darkened pods slightly, but the effects of nitrogen were mixed (Kobayashi et al., 1989). Ascorbic acid content deteriorates with pod color (Akimoto and Kuroda, 1981) and can be used as an index of pod color and freshness.

Edamame releases a sweet aroma when cooked. At the peak of ripeness, aromatic concentrates have much higher levels of (Z)-3-hexenyl acetate, linalool, acetophenone, and cis-jasmone (Sugawara et al., 1988). Several components of beany odor were also identified: hexanal, 1-hexanol, (E)-2-hexenal, 1-octen-3-ol, and 2-pentylfuran. Cis-jasmone is found in edamame only at the peak of ripeness and may be the key component of aroma.

For edamame, the two most important components of flavor are sweet and savory. Its sweet taste is determined by sucrose content and its savory taste probably by amino acids like glutamic acid and alanine (Masuda, Hashizume, and Kameko, 1988). Beany flavor increases with maturity and can be divided into two components–beany and bitter (Rackis et al., 1972). The beany taste may come from linolenic acid oxidized by lipoxygenase and the bitter taste may be the lipoxygenase itself.

Many management choices influence flavor, such as: variety selection, fertilizer application, planting density, harvest procedures, and processing conditions (Masuda, 1989). In Japan, basal fertilizer rates are generally around 50-80 kilograms nitrogen/hectare, 70-100 kilograms phosphorus, and 100-140 kilograms potassium. Excessive basal nitrogen will reduce pod set and increase the number of empty or one-seeded pods. Supplemental nitrogen after anthesis can increase amino-acid levels but has an inconsistent effect on sugars. Kobayashi et al. (1989) found total sugars decreased from 3.18 percent to 2.63-2.94 percent depending on the rate and timing of nitrogen application. Abe et al. (1985) cites contrary data, but adds that pods given supplementary fertilizer may be damaged more easily during processing. Lower plant densities not only darken pods but show consistently higher amino acid levels and sucrose during the maturation period (Chiba, Yaegashi, and Sato, 1989).

The date, timing, and method of harvesting are of crucial importance. Unfortunately, sugar and amino-acid contents peak several days before the beans have adequately filled the pods. When pod color and seed size appear the best, flavor has already begun to deteriorate. In addition, there is a daily cycle for amino acids. A peak occurs near sunset, and the more sunlight during the day, the higher the peak. For glutamic acid, a peak of 255 milligrams gran fresh weight was reported at sunset, with a decline to 160 milli-grams by the next morning at sunrise (Masuda, 1989). After harvest, sugar and amino-acid deterioration can affect flavor within 3-10 hours unless the beans are cooled (Chiba and Yaegashi, 1988), or unless whole plants are harvested. If whole plants are packaged and sealed in low-density, polyethylene bags, they remain acceptable after one week at 20°C (Iwata, Sugiara, and Shirahata, 1982).

If the pods are adequately green, post-harvest handling is the major influence on quality and flavor. The need for rapid cooling cannot be overemphasized. Fresh product often sits at room temperature for 24 hours or more before sale and use. Masuda, Hashizume, and Kaveko (1988) found that frozen edamame from Taiwan often had more sucrose (e.g., 1.7 percent vs. 1.1 percent) and amino acids (e.g., alanine at 30 milligrams/100 gran fresh weight vs. 16 milli-grams) than fresh edamame in Japan. Both were significantly lower than harvest peaks (2.9 percent sucrose and 297-milligrams ala-nine). For frozen edamame, blanching is important to stop the oxidation of fatty acids into undesirable tastes.

Both blanching time and temperature must be controlled to reach a balance between the destruction of trypsin inhibitors and the maintenance of good texture. The length and temperature of storage is also significant. Ascorbic-acid levels decrease 20 percent after six months at –20°C, and 40 percent after one year. For post-harvest management, rapid cooling (or blanching and freezing) is essential to slow enzymatic activity and minimize deterioration.

Yield is influenced by planting density. Higher densities intercept more light, which increases biomass (AVRDC, 1988). Acceptable grade A and B pods/plants decrease with higher densities, while the total yield/area of the pods increases (Shin, 1988). Inter-row spacing influences yield more than intra-row spacing (Kline, 1980).

VARIETY SELECTION

Japanese classify soybeans as summer or fall types (Kono, 1986). Most edamame varieties are temperature-sensitive, summer types; only a few are day-length-sensitive, fall types. Summer types are planted in the spring and harvested immature after 75-100 days, while fall types are planted in early summer and take 105 days or more.

About 170 edamame varieties are listed by the *Nihon Shubyo Kyokai* (JSCA, 1987), and several hundred others are available. Most are determinate (Gotoh, 1984) and can be segregated into approximately ten families with representative types (Kono, 1986; IDA, 1989). Among summer types, Okuhara and Sapporo-midori varieties are in Maturity Group (MG) 0; Osodefuri and Shiroge are in MG I; and Fukura, Mikawashima, and Yukimusume are in MG II. Among fall types, Kinshu, Tsurunoko, and Yuzuru are in MG III. All have white pubescence except Okuhara, Osodefuri, and Shiroge. Special qualities include Fukura's sweetness, Kinshu's dark pods, Mikawashima's numerous three-seeded pods, Osodefuri's good flavor, Shiroge's prolific branching, Tsurunoko's large seeds, and Yukimusume's good pod color after processing. Negative features include Fukura's fragile pods, Mikawashima's viney growth, Okuhara's short harvest period, Sapporo-midori's lack of vigor at low temperatures, and Tsurunoko's tall growth.

Among southern Chinese varieties (Guan, 1977), Sanyuewang, Wuyuewu, and Wuyueba are in MGs I-II; and Baishuiou, Liuyueba, and Baimaoliuyuewang are in MGs III-IV; Jiangyoudou and Daqingdou are fall-type soybeans in MGs V-VI. Sanyuewang, Liuyue-

ba, and Baimaoliuyuewang are short plants. Baishuiou, Baimaoliuyuewang, Jiangyoudou, and Daqingdou are superior in quality,
while Liuyueba, Jiangyoudou, and Daqingdou have high yields.

PRODUCTION

Edamame can be transplanted (*teishoku*) or direct-seeded (*futsu
roji*). In Japan, transplants are used in forced (*sokusei*) and early
(*sojuku*) production systems (Kono, 1986). Planting under forced
conditions in carbon dioxide-enriched, heated greenhouses starts in
November and harvest ends by July. Early-spring production starts
with the planting of seedling nurseries in February; these seedlings
are then transplanted to small plastic tunnels 25-30 days later, with
harvest ending by July. Temperature management is achieved by
placing dark covers over the tunnels during cold periods to increase
temperature and by using ventilation during hot periods (Watanabe,
1988). Field production begins in March and ends by October.
Occasionally, plant tops in seeded fields are cut after the primary
leaf stage to increase branching and pod set (Gotoh, 1984). Early-
summer demand pressures farmers to harvest as early as possible to
obtain higher prices, so the onset of harvest is being continually
advanced through improved crop management and variety development.

Most edamame is harvested by hand. When edamame is sold on
the stem, plants are hand-cut or pulled out with roots intact; unacceptable pods and lower leaves are culled, and branches are tied
together in small, aesthetically pleasing bundles. For sale of harvested pods, plants are cut and pods stripped off, sorted, and packaged. In Japan, electrical-powered, stationary pod strippers are
available, and in Taiwan, AVRDC has developed a manual pod
stripper (AVRDC, 1989). Initial studies on mechanical harvesting
have been conducted in Tennessee (Collins and McCarty, 1969) and
at INTSOY (1987). For frozen product, standard methods for processing have been described (Liu and Shanmugasundaram, 1982).

CURRENT RESEARCH

Quality aspects such as darker green, unblemished pods, earlier
maturity (in Japan), higher pod set (for mechanical harvesting),

lower temperature tolerance, longer harvest period, and a trypsin-inhibitor-free, sweet and savory flavor are major breeding objectives, with adequate seed size, seeds/pod, and pubescence already well established in the breeding lines. Pod shatter is a problem for seed production, but easily opened pods are important for vegetable production. Yield is a secondary consideration because of the high value of edamame, but disease resistance is important in some areas, particularly in the tropics.

Active public-research programs have been founded at AVRDC for variety development (Shanmugasundaram, Tsou, and Cheng, 1989); at INTSOY for mechanical harvesting (INTSOY, 1987); and at Iwate Prefectural Agricultural and Horticultural Experiment Stations for land-race collection (Kiuchi et al., 1987), variety development (Kiuchi et al., 1989), crop management (Chiba, Yaegashi, and Sato, 1989; Kobayashi et al., 1989), and post-harvest management (Chiba and Yaegashi, 1988). Other research programs include the Japanese National Food Research Institute at Tsukuba's study on quality (Masuda, 1989) and at Washington State University's evaluation of varieties and production systems. Many Japanese seed companies (Sakata and Yukijirushi, etc.) also have variety development programs underway. Other prefectures in Japan sponsor local research, and China, Korea, Sri Lanka, and Thailand all have active research programs.

SUMMARY

In the United States, edamame also has potential as an easier-to-grow, better-tasting, more-nutritious substitute for lima beans. Served in the pods, it might appeal to consumers interested in natural foods, particularly if it were grown organically. In spite of drastic changes in the Japanese diet, demand is slowly rising and traditional foods like edamame continue to be very popular (Cook, 1988). In a hungry world, with research and education, edamame could be a nutritious, savory, and inexpensive addition to local diets, especially in calorie-, protein-, and vitamin-deficient regions of the world.

Chapter 16

Developing Trade Relations in Wood Products

Thomas M. Maloney
Roy F. Pellerin

The Wood Materials and Engineering Laboratory at Washington State University has participated actively in IMPACT (International Marketing Program for Agricultural Commodities and Trade) international marketing activities since IMPACT's inception in 1985. University faculty cannot sell any materials directly–this including wood–but they can be influential in helping to develop trade relations. The role of the Wood Materials and Engineering Laboratory (WMEL) in assisting with trade in wood products and equipment from the state of Washington has been established as follows:

1. The development of professional, technical, and industrial contacts in trading countries. These contacts have been established successfully in Japan and are now being opened up in China, Taiwan, and Korea.
2. The exploration of market needs for wood products from the viewpoint of science and technology. For example, the Japanese want clear wood with no knots or blemishes. Therefore, a question to be asked is: How can naturally occurring blemishes be alleviated, or compensated for, so that value is added to the wood in the eyes of the Japanese buyer?
3. The assessment of the potential in exporting U.S. wood technology. The Washington State University (WSU) WMEL has been a pioneer in the development of nondestructive test-

ing, composite wood materials, and wood engineering applications. All of these areas will be valuable as Japan begins to harvest its new-growth timber. Efforts are being made to demonstrate to the Japanese the value of this technology.

4. The appraisal of the trading country's culture and how this culture can be served by new and better wood products.
5. The consideration of how to manufacture or fabricate wood products that will meet a particular country's building codes. For example, Japan has a particularly bewildering combination of prefectural building codes. The WMEL program is finding ways to work with Japanese code agencies in order to gain acceptance of Washington wood products.
6. The ability to analyze the trading country's quality concerns. In Japan, the biggest problems in marketing Pacific Northwest wood products has been one of representative service. British Columbia alone has 20 wood-product representatives in Japan, with an annual budget of about $6 million. The Canadians realize that the Japanese now have about as many housing starts (1.8 million) per year as the United States. The state of Washington only has two representatives in Japan, and these people cover several agricultural products. This limited effort in the case of representative service needs further consideration.
7. The development of a network of scientists, wood technologists, engineers, manufacturers, etc., in both the trading country and in the state of Washington who will need to work together for the advancement of wood material use (Table 16.1). This network is well advanced in the cases of Japan and China (e.g., in developing both the Washington Village concept in Japan and the wood housing project in China), and it is in the initial stages in Taiwan and Korea.

Two major projects are under way at WMEL. These are: "Harmonizing U.S. and Japanese Lumber Grades" and "Developing Wood Housing in the Peoples Republic of China" (reported on separately). However, several other activities have taken place that will also be discussed. These include the network of people and organizations that has been developed; the Forest Products Coordi-

TABLE 16.1

NETWORK OF PEOPLE AND ORGANIZATIONS	
Principal Investigator	Thomas M. Maloney, Washington State University Wood Materials and Engineering Laboratory
Co-investigators	Roy F. Pellerin, Washington State University, Wood Materials and Engineering Laboratory
	Keith A. Blatner, Washington State University, Department of Natural Resource Sciences
	W. Ramsey Smith, University of Washington, Center for International Trade in Forest Products
	Takashi Nakai, Forestry and Forest Products Research Institute, Tsukuba, Japan
	Hideo Sugiyama, Science University, Tokyo, Japan
	He Naizhang, Chinese Academy of Forestry, Beijing, China
	Wang Peiyuan, Chinese Academy of Forestry, Beijing, China
	Li Xiaoming, Chinese Academy of Forestry, Beijing, China
	Cao Jiqiang, Chinese Academy of Forestry, Beijing, China
Extension Associate	Evergreen Partnership
Washington State Departmentof Trade and Economic Development Contact	Kay Eichinger
Washington State Department of Natural Resources Contact	Pat Harper
External Support	Industry associations, private companies, and foundations

nating Board; the new post-laminating plant in Spokane; the Washington Village project; the Technical Advisory Committee to the State Department of Trade and Economic Development (DTED); and various seminars and presentations.

SPOKANE POST-LAMINATING PLANT

In a joint venture with Sumitomo Forestry Co., Ltd., of Japan, the Plum Creek Timber Company built a post-laminating plant in Spokane. The product is primarily for export to Japan. This potential market was noted in the WMEL report on the IMPACT/CINTRA-FOR Technical Visit to Japan in 1985. The plant is designed to produce 45,000 laminated posts per month, and it can also manufacture other laminated products. To make a successful, quality product, the WMEL was engaged to evaluate the fingerjoint quality of the laminates and to produce experimental posts for evaluation here and in Japan. The key was to use the WMEL's nondestructive testing knowledge and instruments (developed at WSU) to fabricate posts meeting the Japanese standards. The research work was successful, and it helped Plum Creek in establishing their joint venture with Sumitomo and in finally building the Spokane plant.

WASHINGTON VILLAGE

A very important project on which WSU is serving as an advisor is the Washington Village in the Hyogo Prefecture in Japan. The Washington Village idea was originally presented by the Evergreen Partnership to Hyogo housing-construction companies. Subsequently, the idea was presented by Washington State Governor Booth Gardner and the Washington Department of Trade and Economic Development to the Hyogo Prefectural Government. The foundation for the idea was based upon the assumption that Washington State (through Governor Gardner) and the Hyogo Prefecture (through Governor Kaihara) would agree upon a cooperative venture to promote trade relations, political relations, and technological transfer. The venture would be entitled "Washington Village" and would involve the construction of approximately 40-50 single-family residences within a "new town" being developed by the Hyogo Housing Supply Corporation. The number of houses was subsequently increased to 171, with the Washington Village theme upgraded to an international village concept.

All Washington Village homes are to reflect the following general guidelines: (1) Western platform construction, known as 2 × 4 construction in Japan; (2) designed jointly by the state of Washington and Japanese architects; (3) built by Hyogo 2 × 4 homebuilders, with the assistance of builders from the state of Washington; and (4) all wood products would be sourced from Washington State suppliers. Additionally, all other building materials necessary to complete construction would also be sourced from Washington.

An agreement to proceed with the Washington Village project was made between Governor Gardner and Governor Kaihara in September 1987. Currently, the new town development is taking place in and around the City of Sanda and is named New Kobe-Sanda International Garden City. Sanda is approximately 40 miles southwest of Kobe and 45 miles from Osaka, an approximate commuting distance of 50 minutes from each city. A one-hour commute is considered average and acceptable in Japan.

The project has progressed to the stage that a model home has been built in Japan. Also, training materials have been developed to train Japanese carpenters.

DTED TECHNICAL ADVISORY COMMITTEE

The DTED has a market study underway to assist the forest products industry in trade. A consultant, Paul Jensen, is doing this study. The plan is to determine which new products could be produced, what technology is available, and finally, to arrange for financing of new plants or plant revisions.

The WMEL's initial role is to have T. M. Maloney and R. F. Pellerin serve on a Technical Advisory Committee along with DTED staff and two industrial people (selected by WSU). The purpose of the committee is to advise on the best technology available or to suggest how to develop the technology necessary for producing the new products.

SEMINARS

The WMEL cooperates with both the Center for International Trade in Forest Products (CINTRAFOR) at the University of Washington and the Evergreen Partnership in putting on special seminars for the forest products industry and associated industries. Two seminars were held in 1989-90: "Forest Products Markets in Korea and Taiwan" and "Europe 1992: Forest Products Markets." At the first seminar, T. M. Maloney, K. Blatner, and J. H. Kwon presented a paper on "Korean Wood Products Market: A Personal View of Korea's Needs."

PRESENTATIONS

As stated earlier, an important role for the WMEL is to develop an extensive network of scientists, laboratories, and industries in other countries. During the year, T. M. Maloney made presentations in Taiwan and Korea entitled "Development of Wood Composites" and "Wood Composites: The Future Hope of The Forest Products Industry," respectively. R. F. Pellerin made presentations in Japan, described later in this report. He also made presentations in Nanjing and Beijing, in the Peoples Republic of China. His work in China was sponsored by the Food and Agricultural Organization (FAO) of the United Nations.

HARMONIZING U. S. AND JAPANESE LUMBER GRADES

There have been many discussions about unfair trade practices in Japan. Some particularly bitter discussions have taken place about wood products. The U.S. Omnibus Trade and Competitiveness Act of 1988 requires that the Office of the United States Trade Representative (USTR) provide reports on the National Trade Estimate (NTE) Report on Foreign Trade Barriers. The April 28, 1989, annual report detailed barriers in Japan. The NTE reported on wood in Japan as follows:

Japan's tariffs on and misclassification of wood products as well as its building codes and product standards that favor other nonwood construction materials continue to dampen the demand for wood in general including many competitive U.S. wood products. The U.S. industry claims that government assistance, fair competition codes, toleration of anticompetitive practices, counterliberalization measures, government procurement policies, and unnecessary restrictive building and fire codes inhibit U.S. exports of wood products to Japan. Japan is the largest U.S. export market for wood and paper products. U.S. exports totaled approximately $3.2 billion in 1988. If Japan's tariffs were reduced to comparable U.S. levels and progress made in modifying Japan's building codes and wood product standards, the potential U.S. trade increase for both wood and paper could be significant. (NFPA, June 1989, p.6)

There are provisions called "Super 301" in the 1988 Trade Act. In dealing with Japan, technical standards were cited first. Later, the agenda was broadened to include tariffs, tariff misclassification, technical standards, and building codes (NFPA, December 1989). In the standards area, the concern has been as follows:

Discriminatory standards and building code restrictions: government restrictions limit the use of wood on exterior and interior construction for both residential and nonresidential building with no technical justification. Japanese test procedures discriminate against proven wood products with results such as a lack of reasonable standards for Oriented Strand Board (OSB), nailbearing standards which discriminate against OSB, prescriptive Japanese standards, and failure to accept internationally-recognized performance standards such as Machine Stress-Rated Lumber. Other examples include obsolete tests and nonuniform application of test results. (NFPA, June 1989, p. 56)

The WMEL effort is designed to help harmonize the Japanese and U.S. systems. Another major effort is to utilize nondestructive testing, so that both Japanese and U.S. wood species are assigned similar strengths and grades.

The most recent effort with Japan has been to nondestructively evaluate the structural properties of their plantation-grown timber.

Wooden members that were 10 cm × 10 cm × 3 m in length (referred to as baby squares) and small, clear specimens (3 cm × 1 cm × 25 cm long) of Sugi (Japanese Cedar) were evaluated nondestructively during the summer of 1989. Destructive testing of these members is being conducted by the Forestry and Forest Products Research Institute in Tsukuba, Japan, under the direction of Takashi Nakai.

Upon completion of the destructive testing, the test results will be forwarded to WSU for analysis. To date, only the data from the tension tests of the baby squares has been received and analyzed. The results of the analysis show a very good relationship between the nondestructive and destructive measurements of modulus of elasticity. The coefficient of determination (or r^2) was 0.96. The results of the analysis of nondestructive and destructive measurements of ultimate tensile strength, however, were much poorer than anticipated. A partial explanation of these findings is that the grips used in the destructive testing of the baby squares were too small for the 10 cm thickness of the tension specimens, thus causing most of the specimens to fail in compression within the grips (instead of within the intended tension failures).

The purpose of this research is to demonstrate the ability to assess the properties of wood nondestructively as well as the subsequent value of the knowledge of these properties for individual pieces in the utilization of the Japanese timber. The scientists with whom we have been working recognize the value of this knowledge and are working at establishing grading methods in Japan that are based on the properties of the timber. With their success in promoting these grading methods, lumber from the United States and, more specifically, the Pacific Northwest should be accepted more readily in Japan.

References

Chapter 2

Cook, Annabel Kirschner. 1988. *The Evolution of Japanese Food Spending Patterns: 1963-1984*. Washington State University: IMPACT Center Report No. 26.

Cook, Annabel Kirschner. 1990. *Patterns of Change in East Asia and Marketing Agricultural Products*. Washington State University: Agricultural Research Bulletin XB1017.

Poleman, T.T. 1981. "Quantifying the Nutrition Situation in Developing Countries." *Food Research Institute Studies*, Vol. 18, No. 1, pp. 1-58.

Population Reference Bureau. 1990. *1990 World Population Data Sheet*. Washington, DC: Population Reference Bureau.

Salathe, Larry. 1979. "The Effects of Changes in Population Characteristics on U.S. Consumption of Selected Foods." *American Journal of Agricultural Economics*, Vol. 61, No. 5, pp. 1038-1045.

Vogel, Suzanne H. 1978. "Professional Housewife: The Career of Urban Middle Class Japanese Women." *The Japan Interpreter*, Vol. 12, No. 1, pp. 16-43.

Chapter 3

Axtell, Roger E. 1989. *The Do's and Taboo's of International Trade*. New York: John Wiley & Sons.

Hall, Edward T. and Mildred Reed Hall. 1987. *Hidden Differences: Doing Business with the Japanese*. Garden City, NY: Anchor/Doubleday.

Chapter 4

Jussaume, Raymond A., Jr. and Annabel Kirschner Cook. 1989. "Japanese Food Consumption Patterns: How Western Are They?" *Pacific Northwest Executive*, Vol. 6, No. 1, pp. 17-20.

Jassaume, Raymond A., Jr., Nobushiro Suzuki and Dean Judson. 1989. *Japanese Wholesale Auction Markets for Fresh Fruits and Vegetables*. Washington State University: IMPACT Center Report No. 36.

Matsumoto, Takeo. 1959. *Staple Food Control in Japan*. Tokyo: Shobi Printing Company.

Ministry of Agriculture, Forestry, and Fisheries. 1987a. *Statistics of Production and Shipment of Vegetables*. Tokyo: Ministry of Agriculture, Forestry, and Fisheries.
Ministry of Agriculture, Forestry, and Fisheries. 1987b. *Reports of Fresh Produce Wholesale Market*. Tokyo: Ministry of Agriculture, Forestry, and Fisheries.
Ministry of Agriculture, Forestry, and Fisheries. 1989. *Oroshiuri Shijyo no Genjyo to Kadai (Wholesale Market Situation)*. Tokyo: Ministry of Agriculture, Forestry, and Fisheries.
Tanaka, Manabu. 1989. "Improvements in the Vegetable Marketing System: Some Suggestions Based on the Japanese Experience." *ASPAC Extension Bulletin*, No. 295, pp. 1-8. Taipei City: Taiwan.
Washiyama, Yuji. 1989. "The Present Marketing Structure for Vegetables, and the Operation of Farmers' Cooperatives in Japan." *ASPAC Extension Bulletin*, No. 295, pp. 9-22. Taipei City: Taiwan.

Chapter 6

Nelson, Donald D., David Youmans, W. Douglas Warnock, and DeVon Knutson. 1990. *The Japanese Beef Market: Implications for Washington Producers*. Pullman, WA: Washington State University Cooperative Extension.
Youmans, David. 1989. "Recent Perceptions of the Pacific Rim Beef Market," (paper) in *Producing for the Japanese Market*, (WSU Beef Information Day, 1989) Moses Lake, April.

Chapter 7

Dean, Bill B., Douglas A. Hasslen, and Jerry McCall. 1988. *An Assessment of Vegetable Production and Handling in Washington*. Washington State University: IMPACT Center Information Series No. 27.
Horticultural Products Review, FAS, January 1990.
Tradescope, April 1989. Japan External Trade Organization (JETRO).
U.S. Seed Exports, FAS, October 1988.

Chapter 10

Australian Bureau of Agricultural and Resource Economics. 1988. *Japanese Agricultural Policies: A Time of Change*. Canberra: Australian Government Publishing Service. Policy Monograph No. 3, Project 11325.
Dick, J.W., K. Shelke, Y. Holm, and K.S. Loo. 1986. "The Effect of Wheat Flour Quality, Formulation, and Processing on Chinese Wet Noodle Quality." *Report for U.S. Wheat Associates*, Department of Cereal Science and Food Technology, North Dakota State University at Fargo, October 21.
Nagao, S. and T. Sato. 1989. Personal Interview in Japan. May 30.
Nagao, S., S. Ishibashi, S. Imai, T. Sato, T. Kanbe, Y. Kaneko, and H. Otsubo. 1979. "Quality Characteristics of Soft Wheat and Their Utilization in Japan

III. Effects of Crop Year and Protein Content on Product Quality." *Cereal Chemistry*, Vol. 54, p. 300.

Pomeranz, Y. 1987. *Modern Cereal Science and Technology*. VCH Publishers. New York.

Statistics Bureau Prime Minister's Office, Japan. 1961-87. *Annual Report on the Family Income and Expenditure Survey*. Various issues, 1961-1987.

Uchida, Michiro. 1988. "Technical Development in Japanese Commercial Baking." The Japan Institute of Baking. Report for 8th International Cereal and Bread Congress. Lausanne, Switzerland.

U.S. Congress/Office of Technology Assessment. 1987. *Enhancing the Quality of U.S. Grain for International Trade*. OTA-F-399. Washington, DC: Government Printing Office, February.

U.S. Congress/Office of Technology Assessment. 1989. *Grain Quality in International Trade: A Comparison of Major U.S. Competitors*. OTA-F-402. Washington, DC: Government Printing Office, February.

Washington Agricultural Statistics. 1989. "Wheat Variety Survey, 1988." Olympia, WA: Washington Agricultural Statistics Service pp. 41-45.

Washington Wheat Commission. 1988. *Foreign Market Development Reports: White Wheat Market Survey*. April, pp. 16-18.

Wilson, William W. 1989. "Differentiation and Implicit Prices in Export Wheat Markets." *Western Journal of Agricultural Economics*, Vol. 14, No. 1, pp. 67-77.

Chapter 11

Endo, S., K. Tanaka, and S. Nagao. 1983. "Implications of Do-Corder Phenomena with Regard to Dough Development." Paper No. 69. AACC 68th Annual Mtg., Kansas City, MO, November.

Endo, S., K. Okada, and S. Nagao. 1985. "Do-Corder Studies on Dough Development III. Mixing Characteristics of Flour Streams and Their Changes During Dough Mixing in Presence of Chemicals." Paper No. 68. AACC 70th Annual Mtg., Orlando, FL, September.

Endo, S., S. Karibe, K. Okada, and S. Nagao. 1988a. "Factors Affecting Gelatinization Properties of Wheat Starch." *Nippon Shokuhin Kogyo Gakkaishi* Vol. 35, No. 7.

Endo, S., K. Okada, S. Nagao, and B. L. D'Appolonia. 1988b. "Comparison of Milling and Analytical Characteristics of Hard Red Spring and Hard Red Winter Wheats." Paper No. 244. AACC 73rd Annual Mtg., San Diego, CA, October.

McMaster, G. J., and H. J. Moss. 1989. "Flour Quality Requirements of Staple Foods of Asia and the Middle East. *ICC '89, Wheat End-Use Properties*. H. Salovaara (ed). Lahti, Finland, pp. 547-553

Nagao, S. 1979. Wheat--Production and Consumption Trends. *Cereal Foods World* Vol. 24, pp. 593-595.

Nagao, S. 1985. "Do-Corder and Its Application in Dough Rheology." Paper No. 26. AACC 70th Annual Mtg. Orlando, FL, September.

Nagao, S. 1988. "Recent Trends of Soft Wheat Uses in Japan." Paper No. 42. AACC 73rd Annual Mtg., San Diego, CA, October.

Nagao, S. 1989. "Advances in Wheat Processing and Utilization in Japan." *Wheat Is Unique.* Y. Pomeranz (ed). St. Paul, MN: American Association of Cereal Chemistry, pp. 607-614.

Nagao, S., S. Imai, T. Sato, Y. Kaneko, and H. Otsubo. 1976. "Quality Characteristics of Soft Wheats and Their Use in Japan. I. Methods of Assessing Wheat Suitability for Japanese Products." *Cereal Chemistry,* Vol. 53, No. 6, pp. 988-997.

Nagao, S., S. Ishibashi, S. Imai, T. Sato, T. Kanbe, Y. Kaneko, and H. Otsubo. 1977. "Quality Characteristics of Soft Wheats and Their Use in Japan. II. Evaluation of Wheats from the United States, Australia, France, and Japan." *Cereal Chemistry,* Vol. 54, No. 1, pp. 198-204.

Noguchi, G., and G. L. Rubenthaler. 1978. "The Role of Flour Constituents of Pacific Northwest Soft Wheats in Japanese Sponge Cake." Paper No. B5-9. Sixth Intern. Cereal and Bread Congress. Winnipeg, Canada: *Cereals 1978 ICC and AACC,* pp. 16-22.

Okada, K. 1989. "Factors Affecting Dough-Mixing Properties." Paper No. 272. AACC 74th Annual Mtg., Washington, DC, November.

Okada, K., Y. Negishi, and S. Nagao. 1984. "Effects of Overgrinding on the Quality Characteristics of Wheat Flour." Paper No. 22. AACC 69th Annual Mtg., Minneapolis, MN, October.

Okada, K., Y. Negishi, and S. Nagao. "Factors Affecting Dough Breakdown During Overmixing." Paper No. 28. AACC 71st Annual Mtg., Toronto, Ont., Canada, October.

Okamoto, T., T. Maruyama, H. Kanematsu, and I. Niiya. 1989a. "Fatty Acid Compositions and Physical Characteristics of Recent Bakery Margarines." *Yukagaku,* Vol. 38, No. 2, pp. 177-183.

Okamoto, T., T. Maruyama, H. Kanematsu, and I. Niiya. 1989b. "Fatty Acid Compositions and Physical Characteristics of Recent Shortenings." *Yukagaku,* Vol. 38, No. 3, pp. 249-255.

Sawabe, T., and M. Uchida. 1978. Japanese Flour Milling Industry." Paper No. B1-9. Sixth Intern. Cereal and Bread Congress. Winnipeg, Canada: *Cereals 1978 ICC and AACC,* pp. 16-22.

Toyokawa, H., G. L. Rubenthaler, J. R. Powers, and E. G. Schanus. 1989a. "Japanese Noodle Qualities. I. Flour Components." *Cereal Chemistry,* Vol. 66, No. 5, pp. 382-386.

Toyokawa, H., G. L. Rubenthaler, J. R. Powers, and E. G. Schanus. 1989b. "Japanese Noodle Qualities. II. Starch Components." *Cereal Chemistry,* Vol. 66, No. 5, pp. 387-391.

Uchida, M. 1979. "The Bread Industry." *Cereal Foods World,* Vol. 24, pp. 596-597.

Uchida, M. 1982. "The Baking Methods and Their Application in Japan." *Proceedings of the 7th World Cereal and Bread Congress.* J. Holes and J. Kratochill (eds.). Elsevier, Amsterdam-Oxford-New York: Elsevier, pp. 697-684.

Chapter 12

Heydon, Richard and A. Desmond O'Rourke. 1981. *Implications of Japanese Economic Growth for Imports of U.S. Agricultural Products with Special Reference to Deciduous Fruit Imports*. Bulletin 0903. College of Agr. Res. Center, Washington State University, Pullman, WA.

Chapter 13

Davidson, R., and J. G. MacKinnon. 1981. "Several Tests for Model Specification in the Presence of Alternative Hypotheses." *Econometrica*, 49:781-793.

Hanson, J. E. 1987. "Bioeconomic Analysis of the Alaskan King Crab Industry." Ph.D. dissertation, Department of Agricultural Economics, Washington State University, Pullman, WA.

Matulich, S. C., J. E. Hanson, and R. C. Mittelhammer. 1988. "A Market Model of the Alaskan King Crab Industry." *NOAA Technical Memorandum*, National Marine Fisheries Service, Seattle, WA.

Chapter 14

Japan Agrinfo Newsletter. 1989. *Imported Beef Stocks Rising*. (December) 7(4): 9.

Japan Agrinfo Newsletter. 1990. *Wagyu Beef Leads in Year-End Supermarket Shopping*. (February) 7(6):7.

Jussaume, Raymond A., Jr. 1989. *Food Consumption in Seattle, Washington and Kobe, Japan*. Washington State University: IMPACT Center Information Series No. 28.

Lin, Biing-Hwan and Hiroshi Mori. 1990. *Expanding Beef Exports to Japan: Background, Opportunities and Strategies*. University of Idaho: Agricultural Experiment Research Series No. 90-1.

Longworth, John W. 1983. *Beef in Japan*. St. Lucia, Australia: University of Queensland Press.

Middaugh, Alan. 1990. "Trade Implications: The U.S. as a Player in the World Meat Trade." Proceedings of the North Asia Beef Symposium. Canberra, Australia (February 1).

Mori, Hiroshi, Biing-Hwan Lin and William D. Gorman. 1989a. *The New U.S.-Japanese Beef Agreement: Some Implications for the U.S. Beef Industry*. University of Idaho: Agricultural Experiment Station Bulletin No. 696.

Mori, Hiroshi, Biing-Hwan Lin and William D. Gorman. 1989b. *Segments of the Japanese Beef Demand: Results of LA/AIDS Analyses and Implications for Trade Liberalization*. Paper Prepared for Presentation at the Annual Meetings of the Japanese Theoretical Economics and Econometrics Society (October): Tsukuba University.

Moss, Charles Derek and Bill Richardson. 1985. "Customer Satisfaction–The Key to Successful and Legally Unfettered Trading. *European Journal of Marketing*. 19(6): 5-11.

Nelson, Donald D., David Youmans, W. Douglas Warnock, and DeVon Knutson. 1990. *The Japanese Beef Market: Implications for Washington Producers.* Pullman, WA: Washington State University Cooperative Extension, EB1567.

Ohga, K. 1988. *Trade Liberalization of Agricultural Products and (its Impacts on) Japanese Agriculture.* Invited paper at the general session of the Japanese Agricultural Economics Society.

Sanderson, F.H. 1987. "United States-Japan Negotiating Issues and Opportunities in the GATT," in D.G. Johnson (ed.), *Agricultural Reform Efforts in the United States and Japan.* New York: New York University Press.

Seng, Philip M. 1989. "Cracking the Japanese Market–Problems and Opportunities." Proceedings of Beef Information Day, Moses Lake, WA. pp. 82-87.

Smith, N. Craig. 1987. "Consumer Boycotts and Consumer Sovereignty." *European Journal of Marketing.* 21(5): 7-19.

Stern, Louis W., and Frederick D. Sturdivant. 1987. "Customer-driven Distribution Systems." *Harvard Business Review.* 65(4):34-41.

Williams, Stephen C. 1986. *Marketing Tuna in Japan.* Queensland, Australia: Queensland Fishing Industry Training Committee.

Chapter 15

Abe, I., Y. Okuda, T. Iwata, and K. Chachin. 1985. "Maintaining Freshness in Machine-Harvested Edamame" ("Kikai Dakkyo Edamame no Sendo Hoji Gijutsu ni Kansuru Kenkyu"). *Engei Gakkai Happyo Yoshi*, Fall S60, pp. 494-495.

Akimoto, K. and S. Kuroda. 1981. "Quality of Green Soybeans Packaged in Perforated PE/PP Film." *Engei Gakkai Zasshi*, Vol. 50, pp. 100-107.

Asian Vegetable Research and Development Center. 1988. Progress Report, Crop Improvement Program, Soybean Physiology. Tainan: AVRDC.

Asian Vegetable Research and Development Center. 1989. Progress Report Summaries 1988. Tainan: AVRDC.

Chiba, Y. and S. Yaegashi. 1988. "Quality Change of Green Soybeans After Harvest." *Tohoku Nogyo Kenkyu*, Vol. 41, pp. 287-288.

Chiba, Y., S. Yaegashi, and H. Sato. 1989. "Cultivation Method of Green Soybean I: Effect of Planting Density on Quality and Yield." *Tohoku Nogyo Kenkyu*, Vol. 42, pp. 279-280.

Collins, J.L. and I.E. McCarty. 1969. "Handling of Vegetable Soybeans Mechanically." *Soybean Digest*, Vol. 12, pp. 20-21.

Cook, A. 1988. *The Evolution of Japanese Food Spending Patterns: 1963-84.* Washington State University: IMPACT Center Report No. 26.

Gotoh, K. 1984. "Historical Review of Soybean Cultivation in Japan." *Tropical Agriculture Research Series*, Vol. 17, pp. 135-142.

Guan, P.Z. 1977. "Vegetable Bean." (*Vegetable Cultivation, South China*). S.X. Li (ed.) Beijing: Chinese Agricultural Publishers, pp. 333-338.

Gupta, A.K., M. Kapoor, and A.D. Deodhar. 1976. "Chemical Composition and Cooking Characteristics of Vegetable and Grain-Type Soybeans." *Journal of Food Science and Technology*, Vol. 13, pp. 133-137.

Haas, P.W., L.C. Gilbert, and A.D. Edwards. 1982. *Fresh Green Soybeans: Analysis of Field Performance and Sensory Qualities.* Emmaus, PA: Rodale Press.

Hong, E.H., S.D. Kim, and Y.H. Hwang. 1984. "Production and Use of and Research on Soybeans in Korea." *Tropical Agriculture Research Series,* Vol. 17, pp. 81-93.

Hymowitz, T. 1984. "Dorsett-Morse Soybean Collection Trip to East Asia: A 50-Year Perspective." *Economic Botany,* Vol. 38, pp. 378-388.

Hymowitz, T., D.W. Mies, and C. J. Klebek. 1971. "Frequency of Trypsin-Inhibitor Variants in Seed Protein of Four Soybean Populations." *East African Agricultural and Forestry Journal,* Vol. 37, pp. 73-77.

Hymowitz, T., F.I. Collins, J. Panczner, and W.M. Walker. 1972. "Relationship Between the Content of Oil, Protein, and Sugar in the Soybean Seed." *Agronomy Journal,* Vol. 64, pp. 613-616.

Igata, S. 1977. *Research on the History of Cereal Crops (Nihon Kodai Kokumotsushi).* Tokyo: Yoshikawa Kobunkan.

International Soybean Program. 1987. "INTSOY Research Focuses on Green Soybeans as Commercial Frozen Vegetable." INTSOY Newsletter, Vol. 37, pp. 1-2.

Iwata, T., H. Sugiura, and K. Shirahata. 1982. "Keeping Quality of Green Soybeans by Whole Plant Packaging." *Engei Gakkai Zasshi,* Vol. 51, pp. 224-230.

Iwate Department of Agriculture (IDA). 1989. *Guide to the Production of Vegetables and Flowers (Yasai, Kaki Saibai Gijutsu Shishin).* Morioka: Iwate-ken.

Iwate Department of Agriculture (IDA). 1990. *Grading and Shipping Standards for Produce (Iwate-ken Seikabutsuto Shukka Kikaku Shido Hikkei).* Morioka: Iwate-ken.

Japan Tariff Association (JTA). 1989. *Japan Exports and Imports.* Tokyo: Japan Tariff Association.

Japanese Seed Company Association (JSCA). 1987. *Vegetable Variety List (Yasai Hinshu Meikan).* Tokyo: Nihon Shubyo Kyokai.

Jian, Y.Y. 1984. "Situation of Soybean Production and Research in China." *Tropical Agriculture Research Series,* Vol. 17, pp. 67-72.

Kiuchi, Y., H. Ishikawa, M. Nitta, and T. Sasaki. 1987. "Collection of Crop Genetic Resources and Their Characteristics in Iwate Prefecture I: Characteristics of Local Soybean Varieties." *Tohoku Nogyo Kenkyu,* Vol. 40, pp. 139-140.

Kiuchi, Y., H. Ishikawa, M. Nitta, T. Sasaki, and T. Sato. 1989. "New Vegetable-Type Soybean Varieties: Iwamamekei 1, Iwamamekei 2, Iwamamekei 3, Iwamamekei 4, and C9." *Iwate Kenritsu Nogyo Shikenjo Kenkyu Hokoku,* Vol. 28, pp. 27-39.

Kline, W.L. 1980. "The Effect of Intra and Interrow Spacing on Yield Components of Vegetable Soybeans." Master's Thesis, Cornell University.

Kobayashi, T., M. Orisaka, K. Miyashita, and Y. Chiba. 1989. "Cultivation Method of Green Soybean 2: Effects of Fertilization and Soil Management on Quality and Yield." *Tohoku Nogyo Kenkyu,* Vol. 42, pp. 281-282.

Kono, S. 1986. "Edamame." *Methods of Bean Production (Sakukei o Ikasu*

Mamerui no   Tsukurikata). Tokyo: Nosangyoson Bunka Kyokai, pp. 195-243.

Lee, D.K. 1986a. "Studies on the Utilization of Green Soybeans 1: Effect of Sowing Dates on Growth and Yield in Green Soybeans." *Research Reports of the Rural Development Administration–Crops (Korea)*, Vol. 28, pp. 136-141.

Lee, D.K. 1986b. "Studies on the Utilization of Green Soybeans 2: Effect of Vinyl Mulching and Tunnels for Early Cultivation on Yield of Green Soybeans. *Research Reports of the Rural Development Administration–Crops (Korea)* Vol. 28, pp. 142-146.

Liener, I.E. 1978. "Nutritional Value of Food Protein Products." *Soybeans: Chemistry and Technology*, A.K. Smith and S.J. Circle (eds.). Westport, CT: AVI, pp. 203-77.

Liu, C. and S. Shanmugasundaram. 1982. "Frozen Vegetable Soybean Industry in Taiwan." *Vegetables and Ornamentals in the Tropics*, M.C. Ali and L.E. Siong (eds.). Serdang: University Pertanian Malaysia, pp. 199-212.

Lloyd, J.W. 1940. "Range and Adaptation of Certain Varieties of Vegetable Soybeans." Illinois Agricultural Experiment Station Bulletin, Vol. 471, pp. 77-100.

Masuda, R. 1989. "Frozen Vegetables–Edamame" ("Yasai no Reito–Edamame"). *Reito*, Vol. 64, pp. 359-376.

Masuda, R., K. Hashizume, and K. Kaneko. 1988. "Effect of Holding Time Before Freezing on the Constituents and the Flavor of Frozen Green Soybeans (Edamame)." *Nihon Shokuhin Kogyo Gakkaishi*, Vol. 35, pp. 763-770.

Ministry of Agriculture, Forestry, and Fisheries (MAFF). 1990. *Statistical Handbook*. Tokyo: Norinsuisansho.

Rackis, J.J. 1978. "Biochemical Changes in Soybeans: Maturation, Post-harvest Storage and Processing, and Germination. *Post-Harvest Biology and Technology*, H.O. Hultin and M. Milner (eds.). Westport: Food and Nutrition, pp. 34-76.

Rackis, J.J. 1981. "Comparison of the Food Value of Immature, Mature, and Germinated Soybeans." *Quality of Fruits and Vegetables in North America*, R. Teranishi and H. Berrero-Benitz (eds.). Washington, DC: American Chemical Society, pp. 183-212.

Rackis, J.J., D.H. Hoing, D.J. Sessa, and H.A. Moser. 1972. "Lipoxygenase and Peroxidase Activities of Soybeans as Related to Flavor Profile During Maturation." *Cereal Chemistry*, Vol. 49, pp. 586-597.

Reddy, P.N., K.N. Reddy, S.K. Rao, and S.P. Singh. 1989. "Effect of Seed Size on Qualitative and Quantitative Traits in Soybean (*Glycine max* (L.) Merrill)." *Seed Science and Technology*, Vol. 17, pp. 289-295.

Shanmugasundaram, S., T.C.S. Tsou, and S.H. Cheng. 1989. "Vegetable Soybeans in the East." *World Soybean Research Conference IV: Actas Proceedings*, A.J. Pascale (ed.), pp. 1978-1987.

Shin, H.R. 1988. "Effect of Drought Stress and Plant Density on Yield and Quality of Vegetable Soybeans." AVRDC Research Report, Tainan.

Shurtleff, W. and A. Aoyagi. Unpublished. "History of Fresh Green Soybeans and

Vegetables-Type Soybeans." *History of Soybeans and Soyfoods*. Lafayette: Soyfoods Center, Chapter 22.

Smith, A.K. and S.J. Circle. 1978. "Chemical Composition of the Seed." *Soybeans: Chemistry and Technology, Volume 1, Proteins*, A.K. Smith and S.J. Circle (eds.). Westport, CT: AVI, pp. 61-92.

Smith, J.M. and F.O. Van Duyne. 1951. "Other Soybean Products." *Soybeans and Soybean Products, Volume 2*, K.S. Markley (ed.). New York: Interscience, pp. 1055-1078.

Sugawara, E., T. Ito, S. Odagiri, K. Kubota, and A. Kobayashi. 1988. "Changes in Aroma Components of Green Soybeans with Maturity." *Nihon Nogei Kagaku Kaishi*, Vol. 62, pp. 148-155.

Sunada, K. 1986. *Methods of Bean Production in Hokkaido (Hokkaido no Mame Saku Gijutsu)*. Sapporo: Nogyo Gijutsu Fukyu Kyokai.

Wang, H.L., G.C. Mustakas, W.J. Wolf, L.C. Wang, C.W. Hesseltine, E.B. Bagley. 1979. "Soybeans as human food–unprocessed and simply processed." *USDA Utilization Report 5*.

Watanabe, M. 1988. "Tunnel Cultivation of Sapporo-midori" ("Tunneru saibai-Sapporo-midori"). *Nogyo Gijutsu Taikei (yasai)*, Vol. 13, pp. 1-8.

Woodruff, S. and H. Klass. 1938. "A Study of Soybean Varieties with Reference to Their Use as Food." Illinois Agricultural Experiment Station Bulletin 443.

Yen, J.T., T. Hymowitz, L. Spilsbury, J.D. Brooks, and A.H. Jensen. 1970. "Utilization by Rats of Protein from Trypsin-Inhibitor Variant Soybeans. *Journal of Animal Science*, Vol. 31 p. 214.

Chapter 16

International Trade Report. 1989. "National Trade Estimate Report Released by USTR." Washington, DC: National Forest Products Association, June, p. 6.

International Trade Report. 1989. "Super 301 Update." Washington, DC: National Forest Products Association, December, p. 2.

International Trade Report. 1989. "NFPA President Optimistic About Super 301 Potential." Washington, DC: National Forest Products Association, June, pp. 5-6.

Index

201